Basics and Applications in Quantum Optics

Basics and Applications in Quantum Optics

Editors

Maria Bondani
Alessia Allevi
Stefano Olivares

MDPI • Basel • Beijing • Wuhan • Barcelona • Belgrade • Manchester • Tokyo • Cluj • Tianjin

Editors
Maria Bondani
National Research Council—Institute for
Photonics and Nanotechnologies
(CNR-IFN)
Italy

Alessia Allevi
University of Insubria
Italy

Stefano Olivares
University of Milan
Italy

Editorial Office
MDPI
St. Alban-Anlage 66
4052 Basel, Switzerland

This is a reprint of articles from the Special Issue published online in the open access journal *Applied Sciences* (ISSN 2076-3417) (available at: https://www.mdpi.com/journal/applsci/special_issues/Basics_and_Applications_in_Quantum_Optics).

For citation purposes, cite each article independently as indicated on the article page online and as indicated below:

LastName, A.A.; LastName, B.B.; LastName, C.C. Article Title. *Journal Name* **Year**, *Volume Number*, Page Range.

ISBN 978-3-0365-3312-4 (Hbk)
ISBN 978-3-0365-3311-7 (PDF)

Cover image courtesy of Alessia Allevi

Contents

About the Editors

Maria Bondani (PhD in Physics in 1994) is a Researcher at the Institute for Photonics and Nanotechnologies of the National Research Council. Her scientific activity covers several topics in the field of radiation–matter interaction, with reference to fluorescence studies, nonlinear optics, quantum optics and light detection techniques in different regimes for the generation and characterization of optical states with applications for quantum information. The research activity is mainly experimental but also includes theoretical aspects (development of models, simulations). She is also involved in physics education research and outreach activities, particularly in the fields of quantum physics and quantum technologies.

Alessia Allevi (Associate Professor) graduated in 2002 and obtained her PhD in Physics in 2006. Since 2019, she has been an Associate Professor at the University of Insubria in Como (Italy). She is an experimentalist, and her main interests are quantum and nonlinear optics, quantum information and the characterization of different classes of photodetectors. Her research activity is supported by several collaborations with national and international theoretical groups.

Stefano Olivares (Associate Professor) received his PhD degree in Physics from the University of Milan, where he is currently Associate Professor at the Department of Physics. He is a theoretician, and his interests include quantum information, quantum estimation, quantum optics, quantum interferometry, and quantum computation. His main contributions are in the fields of quantum estimation of states and operations, the generation and application of entanglement, quantum information, and quantum communication. Although his research activity is mainly theoretical, he is an active collaborator in many experimental groups.

Preface to "Basics and Applications in Quantum Optics"

Quantum technologies are advancing very rapidly and have the potential to innovate communication and computing far beyond current possibilities. Among the possible platforms that are suitable for running quantum technology protocols, in recent decades, quantum optics has received a lot of attention due to the handiness and versatility of optical systems. In addition to studying the fundamentals of quantum mechanics, quantum optical states have been exploited for several applications, such as quantum-state engineering, quantum communication and quantum cryptography protocols, enhanced metrology and sensing, quantum optical integrated circuits, quantum imaging, and quantum biological effects. In this Special Issue, we collect some papers and a review focusing on some recent research activities that show the potential of quantum optics for the advancement of quantum technologies. The addressed topics range from quantum computing to quantum-state engineering, from quantum communication to quantum cryptography, and from quantum simulation to quantum imaging, in perfect agreement with the four pillars of the European Commission Quantum Technologies Flagship Program.

Maria Bondani, Alessia Allevi, Stefano Olivares
Editors

Editorial

Special Issue on Basics and Applications in Quantum Optics

Alessia Allevi [1,2,*,†], Stefano Olivares [3,†] and Maria Bondani [2,†]

[1] Department of Science and High Technology, University of Insubria, Via Valleggio 11, 22100 Como, Italy
[2] Institute for Photonics and Nanotechnologies, CNR, Via Valleggio 11, 22100 Como, Italy; maria.bondani@uninsubria.it
[3] INFN, Milano Section, Department of Physics "Aldo Pontremoli", University of Milan, Via Celoria 16, 20133 Milano, Italy; stefano.olivares@fisica.unimi.it
* Correspondence: alessia.allevi@uninsubria.it; Tel.: +39-031-238-6253
† These authors contributed equally to this work.

1. Introduction

Quantum technologies are advancing very rapidly and have the potential to innovate communication and computing far beyond current possibilities. Among the possible platforms suitable to run quantum technology protocols, in the last decades quantum optics has received a lot of attention for the handiness and versatility of optical systems. In addition to studying the fundamentals of quantum mechanics, quantum optical states have been exploited for several applications, such as quantum-state engineering, quantum communication and quantum cryptography protocols, enhanced metrology and sensing, quantum optical integrated circuits, quantum imaging, and quantum biological effects. In this Special Issue, we collect some papers and also a review on some recent research activities that show the potential of quantum optics for the advancement of quantum technologies.

2. Quantum Optics Applications

The topics addressed in the Special Issue range from quantum computing to quantum-state engineering, from quantum communication to quantum cryptography, from quantum simulation to quantum imaging, in perfect agreement with the four pillars of the European Commission Quantum Technologies Flagship Program.

The first paper [1] of this Special Issue, authored by A. Candeloro et al., focuses on an enhanced version of an all-optical system used to implement a quantum finite automaton [2]. The considered automaton recognizes a well-known family of unary periodic languages that play a crucial role in Descriptional Complexity Theory and in the area of Formal Language Theory. The performance of the device benefits from considering the orthogonal output polarizations of the employed single photons. Moreover, the effect of the detector dark counts on the proper operation of the automaton is taken into account. This kind of photonic quantum automaton could be hardwired into "hybrid" architectures that combine classical and quantum components to build very succinct finite-state devices operating in environments where dimension and energy absorption are particularly critical issues.

The paper written by G. Chesi et al. addresses the topic of quantum-state engineering. The authors present the generation and characterization of Sub-Poissonian states by means of conditional measurements performed on multi-mode twin-beam states [3]. These measurements are based on the use of Silicon photomultipliers [4], a class of photon-number-resolving detectors. Such detectors, very compact and cheap, can open new perspectives in the field of quantum optics and quantum technologies, being suitable for investigating mesoscopic states of light [5]. In the paper, a comprehensive model taking into account all the features of the employed detectors is developed and experimentally verified.

In their paper, M.-S. Kang et al. develop a quantum message authentication protocol for improving security against an existential forgery by means of single-qubit unitary operations [6]. The protocol consists of two parts: a quantum encryption and a correspondence

Citation: Allevi, A.; Olivares, S.; Bondani, M. Special Issue on Basics and Applications in Quantum Optics. *Appl. Sci.* **2021**, *11*, 10028. https://doi.org/10.3390/app112110028

Received: 15 October 2021
Accepted: 22 October 2021
Published: 26 October 2021

Publisher's Note: MDPI stays neutral with regard to jurisdictional claims in published maps and institutional affiliations.

Appl. Sci. **2021**, *11*, 10028. https://doi.org/10.3390/app112110028   https://www.mdpi.com/journal/applsci

check. The first part is realized by means of a linear combination of wave plates [7], while the second one is performed using the Hong–Ou–Mandel interference [8]. The successful experimental implementation of the protocol proves that the employed optical system can be considered as the base technology for a complete quantum cryptosystem providing confidentiality, authentication, integrity, and nonrepudiation.

Furthermore, the paper written by K. Park et al. is devoted to quantum-state engineering [9]. Starting form the recent proposal of obtaining high-purity bi-photon states without degrading brightness and collection efficiency by means of a nonlinear interferometer [10], the authors experimentally investigate the fine tunability of the nonlinear interference method to match constructive interference patterns, while maintaining the high spectral purity of the biphoton state. Their results enrich the usefulness and practicality of the method based on the nonlinear interferometer for the efficient generation of photon pairs with high spectral purity, which represents an excellent practical source for quantum information protocols.

The paper authored by A. Allevi et al. focuses on the role of losses in the degradation of the nonclassicality of mesoscopic quantum states of light to be used for secure data transmission in quantum communication protocols [11]. In particular, the authors investigate, both theoretically and experimentally, the effect caused by two realistic kinds of statistically-distributed amounts of loss, namely a Gaussian distribution and a log-normal one, on the nonclassical photon-number correlations between the two parties of multi-mode twin-beam states [12]. The achieved results show to what extent the involved parameters, both those connected to loss and those describing the employed states of light, preserve nonclassicality.

In the last research paper, J. Liñares et al. present the physical simulation of the dynamical and topological properties of atom-field quantum interacting systems by means of integrated quantum photonic devices [13]. The photonic device consists of integrated optical waveguides supporting two collinear modes, which are coupled by integrated optical gratings [14]. The two-mode photonic device with a single-photon quantum state represents the quantum system, and the optical grating corresponds to an external field. This photonic simulator can be regarded as a basic brick for constructing more complex photonic simulators.

Finally, in the review paper by C. Abbattista et al. the advancement of the research toward the design and implementation of quantum plenoptic cameras is presented and discussed [15]. At variance with standard plenoptic cameras, these devices have dramatically-enhanced features, such as diffraction-limited resolution, large depth of focus, and ultra-low noise [16]. For the quantum advantages of the proposed devices to be effective and appealing to end-users, the authors propose to develop high-resolution single-photon avalanche photodiode arrays and high-performance low-level programming of ultra-fast electronics, combined with compressive sensing and quantum tomography algorithms, with the aim of reducing both the acquisition and the elaboration time by two orders of magnitude. These new strategies will open the way to new opportunities and applications, such as for biomedical imaging, security, space imaging, and industrial inspection.

Author Contributions: Conceptualization, A.A., S.O. and M.B.; methodology, A.A.; writing—original draft preparation, A.A., S.O. and M.B; writing—review and editing, A.A., S.O. and M.B. All authors have read and agreed to the published version of the manuscript.

Funding: This research received no external funding.

Institutional Review Board Statement: Not applicable.

Informed Consent Statement: Not applicable.

Data Availability Statement: The datasets used and analysed during the current study are available from the corresponding author on reasonable request.

Acknowledgments: This issue would not have been possible without the contributions of several valued authors, professional reviewers, and the Applied Sciences editorial team. We first extend our congratulations to all the authors. Second, we would like to take this opportunity to show our sincere gratitude to all reviewers. Finally, we thank the editorial team of Applied Sciences and especially Patrick Han.

Conflicts of Interest: The authors declare no conflict of interest.

References

1. Candeloro, A.; Mereghetti, C.; Palano, B.; Cialdi, S.; Paris, M.G.A.; Olivares, S. An enhanced photonic quantum finite automaton. *Appl. Sci.* **2021**, *11*, 8768. [CrossRef]
2. Mereghetti, C.; Palano, B.; Cialdi, S.; Vento, V.; Paris, M.G.A.; Olivares, S. Photonic realization of a quantum finite automaton. *Phys. Rev. Res.* **2020**, *2*, 013089. [CrossRef]
3. Chesi, G.; Allevi, A.; Bondani, M. Conditional Measurements with Silicon Photomultipliers. *Appl. Sci.* **2021**, *11*, 4579. [CrossRef]
4. Akindinov, A.V.; Martemianov, A.N.; Polozov, P.A.; Golovin, V.M.; Grigoriev, E.A. New results on MRS APDs. *Nucl. Instrum. Methods Phys. Res. Sect. A* **1997**, *387*, 231–234. [CrossRef]
5. Cassina, S.; Allevi, A.; Mascagna, V.; Prest, M.; Vallazza, E.; Bondani, M. Exploiting the wide dynamic range of silicon photomultipliers for quantum optics applications. *EPJ Quantum Technol.* **2021**, *8*, 4. [CrossRef]
6. Kang, M.-S.; Kim, Y.-S.; Choi, J.-W.; Yang, H.-J.; Han, S.-W. Experimental Quantum Message Authentication with Single Qubit Unitary Operation. *Appl. Sci.* **2021**, *11*, 2653. [CrossRef]
7. Clarke, R.B.M.; Kendon, V.M.; Chefles, A.; Barnett, S.M.; Riis, E.; Sasaki, M. Experimental realization of optimal detection strategies for overcomplete states. *Phys. Rev. A* **2001**, *64*, 012303. [CrossRef]
8. Horn, R.T.; Babichev, S.; Marzlin, K.-P.; Lvovsky, A.; Sanders, B.C. Single-qubit optical quantum fingerprinting. *Phys. Rev. Lett.* **2005**, *95*, 150502. [CrossRef] [PubMed]
9. Park, K.; Lee, D.; Shin, H. Tunability of the Nonlinear Interferometer Method for Anchoring Constructive Interference Patterns on the ITU-T Grid. *Appl. Sci.* **2021**, *11*, 1429. [CrossRef]
10. Cui, L.; Su, J.; Li, J.; Liu, Y.; Li, X.; Ou, Z.Y. Quantum state engineering by nonlinear quantum interference. *Phys. Rev. A* **2020**, *102*, 033718. [CrossRef]
11. Allevi, A.; Bondani, M. Tailoring Asymmetric Lossy Channels to Test the Robustness of Mesoscopic Quantum States of Light. *Appl. Sci.* **2020**, *10*, 9094. [CrossRef]
12. Allevi, A.; Bondani, M. Preserving nonclassical correlations in strongly unbalanced conditions. *J. Opt. Soc. Am. B* **2019**, *36*, 3275–3281. [CrossRef]
13. Liñares, J.; Prieto-Blanco, X.; Carral, G.M.; Nistal, M.C. Quantum Photonic Simulation of Spin-Magnetic Field Coupling and Atom-Optical Field Interaction. *Appl. Sci.* **2020**, *10*, 8850. [CrossRef]
14. Lee, D. *Electromagnetic Principles of Integrated Optics*; Wiley: New York, NY, USA, 1986.
15. Abbattista, C.; Amoruso, L.; Burri, S.; Charbon, E.; Di Lena, F.; Garuccio, A.; Giannella, D.; Hradil, Z.; Iacobellis, M.; Massaro, G.; et al. Towards Quantum 3D Imaging Devices. *Appl. Sci.* **2021**, *11*, 6414. [CrossRef]
16. Pepe, F.V.; Di Lena, F.; Mazzilli, A.; Garuccio, A.; Scarcelli, G.; D'Angelo, M. Diffraction-limited plenoptic imaging with correlated light. *Phys. Rev. Lett.* **2017**, *119*, 243602. [CrossRef] [PubMed]

applied sciences

Article

An Enhanced Photonic Quantum Finite Automaton

Alessandro Candeloro [1,†], Carlo Mereghetti [1,†], Beatrice Palano [2,†], Simone Cialdi [1,3,†], Matteo G. A. Paris [1,3,†] and Stefano Olivares [1,3,*,†]

1   Dipartimento di Fisica "Aldo Pontremoli", Università degli Studi di Milano, via Celoria 16, I-20133 Milano, Italy; alessandro.candeloro@unimi.it (A.C.); mereghetti@di.unimi.it (C.M.); simone.cialdi@mi.infn.it (S.C.); matteo.paris@fisica.unimi.it (M.G.A.P.)
2   Dipartimento di Informatica "Giovanni Degli Antoni", Università degli Studi di Milano, via Celoria 18, I-20133 Milano, Italy; palano@di.unimi.it
3   INFN, Sezione di Milano, via Celoria 16, I-20133 Milano, Italy
*   Correspondence: stefano.olivares@fisica.unimi.it
†   These authors contributed equally to this work.

Abstract: In a recent paper we have described an optical implementation of a *measure-once one-way quantum finite automaton* recognizing a well-known family of unary periodic languages, accepting words not in the language with a given error probability. To process input words, the automaton exploits the degree of polarization of single photons and, to reduce the acceptance error probability, a technique of confidence amplification using the photon counts is implemented. In this paper, we show that the performance of this automaton may be further improved by using strategies that suitably consider *both* the orthogonal output polarizations of the photon. In our analysis, we also take into account how detector dark counts may affect the performance of the automaton.

Keywords: quantum finite automata; periodic languages; confidence amplification; photodetection

Citation: Candeloro, A.; Mereghetti, C.; Palano, B.; Cialdi, S.; Paris, M.G.A.; Olivares, S. An Enhanced Photonic Quantum Finite Automaton. *Appl. Sci.* **2021**, *11*, 8768. https://doi.org/10.3390/app11188768

Academic Editor: Caterina Ciminelli

Received: 10 August 2021
Accepted: 18 September 2021
Published: 21 September 2021

Publisher's Note: MDPI stays neutral with regard to jurisdictional claims in published maps and institutional affiliations.

1. Introduction

In the recent years, quantum computers have eventually leaped out of the laboratories [1] and become accessible to a still growing community interested in investigating their actual potentialities. Nevertheless, a full-featured quantum computer is still far from being built. However, it is reasonable to think of classical computers exploiting some quantum components. In this framework, quantum finite automata [2,3]—theoretical models for quantum machines with finite memory—may play a key role, as they model small-size quantum computational devices that can be embedded in classical ones. Among possible models, the so-called measure-once one-way quantum finite automaton [4,5] is the simplest, and it has been shown to be the most promising for a physical realization [6]. In fact, restricted models of computation, such as quantum versions of finite automata, have been theoretically studied [7–9] and, very recently, experimentally investigated [6,10].

In [6], a measure-once one-way quantum finite automaton recognizing a well-known family of unary periodic languages [4], namely, languages L_m, has been implemented using quantum optical technology [11,12]. In our implementation, a given input word is accepted by the automaton, with a given error probability, whenever a single photon arrives at the output of the device with a specific polarization. In particular, the experimental realization, based on the manipulation of single-photon polarization and photodetection, has demonstrated the possibility of building small quantum computational component to be embedded in more sophisticated and precise quantum finite automata or also in other computational systems and approaches [13–15]. Albeit the photonic automaton realized in [6] is fed with single photons, it works in a regime where polarized laser pulses (coherent states) are enough, up to detecting the intensity of the output signals instead of counting the number of photons successfully passing through the device with a given polarization (see in [6] for details).

In this paper, we propose an enhanced version of our photonic automaton mentioned above, where, to further reduce the acceptance error probability, we consider not only the photons with the "correct" polarization, but also the other ones. To achieve this goal, the use of single-photon techniques turns out to be crucial, such as the detection of coincidence count to reduce the dark-count rate of the photodetectors [16]. Analytical and numerical results, supported by simulated experiments, show that the enhanced version allows to reduce the error probability by orders of magnitude compared to the previous version, or, analogously, to drastically reduce the mean number of photons needed to achieve the same performance.

The paper is structured as follows. As our work requires some previous knowledge from Theoretical Computer Science about formal languages and finite automata, Section 2 is devoted to introduce the reader to these topics, providing the relevant motivations. In Section 3, we briefly review basics of formal language theory and the definition of a measure-once one-way quantum finite automaton. Section 4 describes the implementation of the measure-once one-way quantum finite automaton based on the polarization of single photons, linear optical elements, and photodetectors. In Section 5, we explain how to improve the confidence of the obtained measure-once one-way quantum finite automaton by processing the number of counts at the detectors. We also introduce new strategies that reduce the error probability, namely, the probability that a "wrong" word is accepted by the automaton or a "correct" word is rejected. The numerical results and the simulated experiments are reported in Section 6. We close the paper with some concluding remarks in Section 7.

2. Formal Languages, Finite Automata, and Quantum Computing

In this section, we would like to expand on motivations that have been driving our research covered by the present contribution and the previous one in [6]. The aim of our work, that bridges between Theoretical Computer Science and Experimental Quantum Optics, has been and is to show that a quantum computing device with finite memory is physically realizable by means of photonics, using a very limited amount of "quantum hardware". To the best of our knowledge, our physical implementation, described here and in [6], of a quantum finite automaton for language acceptation is the first proposed in the literature. Thus, we have shown how the quantum behaviour of microscopic systems can actually represent a computational resource, as theoretically established within the discipline of Quantum Computing. From this perspective, the simple language L_m, introduced in the next section and for which we build our photonic quantum finite automaton, is not really the point here. Instead, the point is the concrete creation of a programmable fully quantum computer with finite memory.

With this being said, we would also like to quickly comment on the language L_m from a Theoretical Computer Science viewpoint. Notwithstanding its simplicity, the language L_m plays a crucial role in Descriptional Complexity Theory (see, e.g., in [17–21]), the area of Formal Language Theory in which the *size* of computational models is investigated. In particular, a well-consolidated trend in Descriptional Complexity is devoted to study the size of several types of *finite automata*. The reader is referred to, e.g., the work in [22] for extensive presentations of automata theory. Very roughly speaking, the hardware of a (one-way) finite automaton A features a read-only input tape consisting of a sequence of cells, each one being able to store an input symbol. The tape is scanned by an input head always moving one position right at each step. At each time during the computation of A, a finite state control is in a state from a *finite* set Q. Some of the states in Q are designated as accepting states, while a state $q_0 \in Q$ is a designated initial state. The computation of A on a word (i.e., a finite sequence of symbols) ω from a given input alphabet begins by having (i) ω stored symbol by symbol, left to right, in the cells of the input tape; (ii) the input head scanning the leftmost tape cell; and (iii) the finite state control being in the state q_0. In a move, A reads the symbol below the input head and, depending on such a symbol and the state of the finite state control, it switches to the next state according to a fixed transition function and moves the input head one position forward. We say that A accepts ω whenever it enters an accepting state after scanning the rightmost symbol of ω; otherwise, A rejects ω. The language accepted by A consists of all the input words accepted by A.

The one described so far is the original model of a finite automaton, called *deterministic*. Several variants of such an original model have been introduced and studied in the literature, sharing the same hardware but different dynamics. Therefore, we have *nondeterministic*, *probabilistic* and, recently, *quantum* finite automata (see, e.g., in [23–25]). Furthermore, *two-way* automata are studied, where the input head can move back and forth on the input tape.

Finite automata represent a formidable theoretical model used in the design and analysis of several devices such as the control units for vending machines, elevators, traffic lights, combination locks, etc. Particularly important is the use of finite automata in very large-scale integration (VLSI) design, namely, in the project of sequential networks which are the building blocks of modern computers and digital systems. Very roughly speaking, a sequential network is a boolean circuit equipped with memory. Engineering a sequential network typically requires modeling its behaviour with a finite automaton whose number of states directly influences the amount of hardware (i.e., the number of logic gates) employed in the electronic realization of the sequential network. From this point of view, having fewer states in the modeling finite automaton directly results in employing smaller hardware which, in turn, means having less energy absorption and fewer cooling problems. These "physical" considerations, that are of paramount importance given the current level of digital device miniaturization, have led to define the size of a finite automaton as the number of its states. In particular, reducing or increasing the number of states is studied, when using different computational paradigms (e.g., deterministic, nondeterministic, probabilistic, quantum, one-way, and two-way) on a finite automaton to perform a given task. Here, is where our simple language L_m comes into play. In fact, this language is universally used as a benchmark to emphasize the succinctness of several types of automata. Several results in the literature shows that accepting L_m on classical models of finite state automata is particularly size-consuming (i.e., it requires a great number of states), while only two basis states are enough on quantum finite automata, as we will see in the next section.

Modular design frameworks have been theoretically proposed [7–9], where more reliable and sophisticated quantum automata can be built by suitably composing (see, e.g., in [26]) easy-to-obtain variants of the quantum automaton for L_m. Hence, our work provides crucial and concrete quantum components for such frameworks, and suggest investigating a physical implementation of some automata composition laws. More generally, the Krohn–Rhodes decomposition theorem [27] states that any classical finite automaton can be simulated by composing very "simple" finite automata: one of these simple automata is exactly the one for L_m. From this perspective, our photonic quantum automaton could be hardwired into "hybrid" architectures joining classical and quantum components to build very succinct finite state devices operating in environments where dimension and energy absorption are particularly critical issues (e.g., drone or robot-based systems [28]).

3. Measure-Once One-Way Quantum Finite Automaton

Here, we briefly overview the main concepts on automata and formal language theory. We refer the interested reader to any of the standard books on these subjects (see, e.g., in [22]), as well as to our contribution [6].

An alphabet is any finite set Σ of elements called symbols. A word on Σ is any sequence $\sigma_1 \sigma_2 \cdots \sigma_n$ with $\sigma_i \in \Sigma$. The set of all words on Σ is denoted by Σ^*. A language L on Σ is any subset of Σ^*, i.e., $L \subseteq \Sigma^*$. If $|\Sigma| = 1$, we say that Σ is a *unary* alphabet, and languages on unary alphabets are called unary languages. In case of unary alphabets, we customarily let $\Sigma = \{a\}$ so that a unary language is any set $L \subseteq a^*$. We let a^k be the unary word obtained by concatenating k times the symbol a.

In what follows, we will be interested in the unary language L_m defined as

$$L_m = \{a^k \mid k \in \mathbb{N} \text{ and } k(\mathrm{mod}\ m) = 0\}. \tag{1}$$

This language is rather famous in the realm of automata theory, since it has proven particularly "size-consuming" to be accepted by several models of classical automata, as the number of needed states increases with m [6]. The reader may find a deep investigation on this fact in the literature [22,29–31]. On the other hand, as presented in [6], very succinct measure-once one-way quantum finite automata (1qfa's, from now on) may be designed and physically realized for L_m. Let us now sketch the main ingredients for a 1qfa accepting L_m.

If we consider the two orthogonal states $|H\rangle = (1,0)$ and $|V\rangle = (0,1)$, the 1qfa is defined as (here we use the formalism based on the Dirac's notation; the analysis based on a more general formalism can be found in [6])

$$\mathcal{A}_1 = \left\{ |H\rangle, U_m, P^H \right\} \tag{2}$$

where $|H\rangle$ represents the initial state, the unitary operation applied by the automaton upon processing any input symbol a is defined as

$$U_m = \exp(-i\theta_m \sigma_y) \tag{3}$$

$$= \begin{pmatrix} \cos\theta_m & \sin\theta_m \\ -\sin\theta_m & \cos\theta_m \end{pmatrix}. \tag{4}$$

with $\theta_m = \pi/m$ and σ_y the Pauli matrix, while $P^H = |H\rangle\langle H|$ is the projector onto the mono-dimensional accepting subspace spanned by $|H\rangle$. The probability $p_{\mathcal{A}_1}(a^k)$ that the 1qfa $\mathcal{A}_1$ accepts the word a^k writes as

$$p_{\mathcal{A}_1}(a^k) = p^H(a^k) \equiv |\langle H|U_m^k|H\rangle|^2 \tag{5}$$

$$= \cos^2(k\theta_m) \rightarrow \begin{cases} = 1 & k(\mathrm{mod}\ m) = 0 \\ \leq \cos^2\theta_m & \text{otherwise.} \end{cases} \tag{6}$$

Therefore, the 1qfa $\mathcal{A}_1$ perfectly recognizes the word $a^k \in L_m$, as we can set a *cut point* λ and an *isolation* ρ to the following values (see in [6] for details on accepting languages with isolated cut point)

$$\lambda = \frac{1 + \cos^2\theta_m}{2} \quad \text{and} \quad \rho = \frac{1 - \cos^2\theta_m}{2}. \tag{7}$$

However, $\mathcal{A}_1$ may also recognize an input word not in L_m with a non-null probability. In the following, we let a^{k_1} with $k_1(\mathrm{mod}\ m) = 1$ any of the word with the highest probability of erroneously being accepted, i.e., $\cos^2\theta_m$, which tends to 1 as m gets large. This can be seen also by the fact that $\rho \rightarrow 0$ as m increases.

As matter of fact, we can also introduce the following 1qft, where we still consider the initial state $|H\rangle$, but, at the output, we focus on the final projection involving the state $|V\rangle$, namely

$$\mathcal{A}_2 = \left\{ |H\rangle, U_m, \mathbb{I} - P^V \right\}, \tag{8}$$

where $P^V = |V\rangle\langle V|$. Indeed, $\mathcal{A}_2$ is formally equivalent to $\mathcal{A}_1$, as $\mathbb{I} - P^V \equiv P^H$. In fact, the probability of accepting a word is now given by

$$p_{\mathcal{A}_2}(a^k) = 1 - p^V(a^k) \rightarrow \begin{cases} = 1 & k(\mathrm{mod}\ m) = 0 \\ \leq \cos^2\theta_m & \text{otherwise} \end{cases} \tag{9}$$

that is the same as in Equation (6), as one may expect. Nevertheless, we show in the next section that the two are not equivalent in a photonic implementation for reasons that will be clear soon.

4. Photonic Implementation of the 1qfa

A photonic implementation of the 1qfa described in the previous section was proposed and demonstrated in [6]. Figure 1 depicts the main elements of the enhanced version of the automata we will describe in the following.

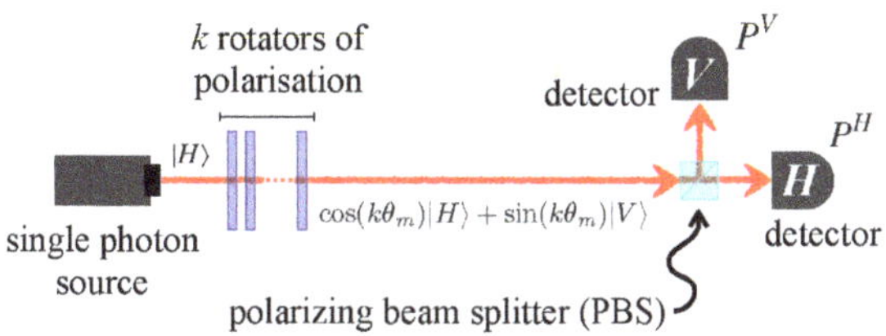

Figure 1. Scheme of the photonic implementation of the 1qfa highlighting the main involved optical elements. See the text for details.

The state of the automaton is encoded in the polarization of single photons, and the Hilbert space is $\mathcal{H} = \mathrm{span}\{|H\rangle, |V\rangle\}$. A single photon source generates a horizontal-polarized state, $|H\rangle$, which is sent to k rotators of polarization, a^k being the input word to be processed. Each rotator corresponds to a unitary rotation of an amount θ_m, which is thus language-dependent. After the rotators, the single photon state reads

$$|k\theta_m\rangle = \cos(k\theta_m)|H\rangle + \sin(k\theta_m)|V\rangle \tag{10}$$

and it is sent to a polarizing beam splitter (PBS; see Figure 1) that reflects the vertical polarization component and transmits the horizontal one. Finally, two photodetectors placed after the PBS realize the projective measure of P^H and P^V. As the reader can see, the scheme is almost the same of that proposed in [6], but here we will implement a new inference strategy exploiting the outcomes from both the detectors.

As we observed in the previous section, the automata $\mathcal{A}_1$ and $\mathcal{A}_2$ accept with certainty a word a^k that belongs to L_m. However, there is a high probability that an incorrect word, such as a^{k_1} with $k_1 \mod m \neq 0$ can be accepted, as we can see from Equations (6) and (9). Therefore, strategies based on a single-photon shot may not be the optimal way to recognize an arbitrary word a^k.

5. Confidence Amplification: An Enhanced Strategy

To reduce the probability of error, we can adopt a technique of confidence amplification as also proposed in [6], namely, we sent a mean number of photons $\langle N_c \rangle$ and we count the number of click $N_c^x(k)$ at the photodetector $x = H, V$, see Figure 1. Therefore, the observed detection frequency at detector $x = H, V$ for an input word a^k will be

$$f_k^x = \frac{N_c^x(k)}{\langle N_c \rangle} \xrightarrow{\langle N_c \rangle \gg 1} p_{\mathcal{A}_{i=1,2}}^x(a^k) \cdot . \tag{11}$$

Thereafter, we turn our problem into that of discriminating among the corresponding detection frequencies and, in particular, we can focus on those related to $k = 0$ (or, equivalently, $k \mod m = 0$) and $k = 1$ (or, more in general, $k \mod m = 1$), since if $k > 1$ one has $f_k^H < f_1^H$ ($f_k^V > f_1^V$). To implement this strategy, we set a threshold frequency as

$$f_{\text{th}}^x = \frac{f_0^x + f_1^x}{2} = \begin{cases} \dfrac{1 + f_1^H}{2} & x = H, \\[2ex] \dfrac{f_1^V}{2} & x = V, \end{cases} \tag{12}$$

where f_1^H (f_1^V) is the highest (lowest) frequency of erroneously accepted words a^{k_1}, while f_0^H (f_0^V) is the frequency corresponding to the correct word. In this formula, we have distinguished the two different strategies: for the H detector, $f_0^H = 1$, as the corresponding photon will always be detected; instead, for the V detector, $f_0^V = 0$, as no photon is detected when the word belongs to L_m. Therefore, the strategy is to accept the word if $f_k^H > f_{\text{th}}^H$ ($f_k^V < f_{\text{th}}^V$) and reject it if $f_k^H < f_{\text{th}}^H$ ($f_k^V > f_{\text{th}}^V$). From now on, we will refer to these strategies as "H strategy" and "V strategy", respectively.

In an ideal scenario, namely, without fluctuations in the sent number of photons, it is clear that the two approaches are complementary and yield to the same conclusion, as the single detections in H and V are perfectly correlated. Moreover, given that only the words $a^k \in L_m$ satisfy the condition $f_k > f_{\text{th}}$, with this strategy we have a zero error probability, provided that $\langle N_c \rangle$ is large enough such that the integer part of N_{th}^H (N_{th}^V) is strictly positive (negative) than $N_c^H(k_1)$ ($N_c^V(k_1)$), i.e., we have the conditions

$$\left\lfloor N_{\text{th}}^H \right\rfloor = \left\lfloor \frac{\langle N_c \rangle (1 + \cos^2 \theta_m)}{2} \right\rfloor > \left\lfloor \langle N_c \rangle \cos^2 \theta_m \right\rfloor, \tag{13}$$

$$\left\lfloor N_{\text{th}}^V \right\rfloor = \left\lfloor \frac{\langle N_c \rangle \sin^2 \theta_m}{2} \right\rfloor < \left\lfloor \langle N_c \rangle \sin^2 \theta_m \right\rfloor. \tag{14}$$

In Figure 2 (black lines and dots), we report the minimum values of $\langle N_c \rangle$ such that the last two inequalities hold.

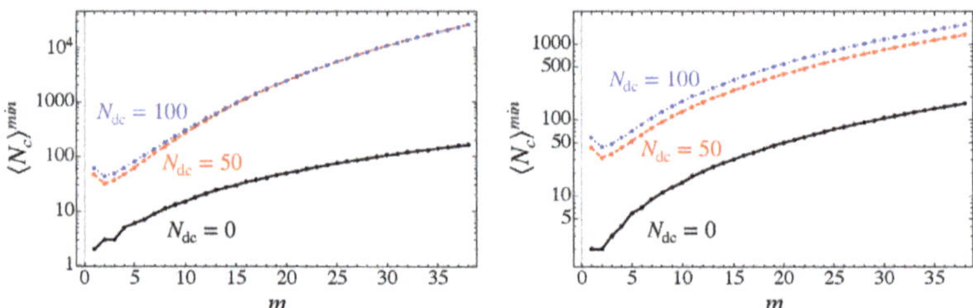

Figure 2. Black line and dots: minimum value $\langle N_c \rangle^{min}$ such that Equation (13) (**left plot**) and Equation (14) (**right plot**) are satisfied as a function of m in the absence of dark counts ($N_{\text{dc}} = 0$). Red line and dots ($N_{\text{dc}} = 50$), blue line and dot ($N_{\text{dc}} = 100$): minimum vale $\langle N_c \rangle^{min}$ such that Equation (22) (**left plot**) and Equation (23) (**right plot**) are satisfied. Notice the different scaling for the y-axis.

In a realistic scenario, the photo-detection is influenced by two distinct noisy effects that can affect the error probability. The first is that the number of detected photons follows a Poisson distribution [32], that is, we have

$$\text{Poi}(n; \mu) = \frac{\mu^n e^{-\mu}}{n!} \tag{15}$$

that is, the probability of detect n photons depends on the average number of detected photons μ. How this affects the error probability has been thoroughly addressed both theoretically and experimentally in [6].

The second effect that we should consider in order to apply our enhanced strategy is due to the dark counts, namely, the random counts registered by the detector without any incident light on it. Being still related to the detection process, also the dark counts follow a Poissonian distribution, whose mean $\langle N_{dc} \rangle$ depends on the particular detector one choose to use. In a typical quantum optics experiment, the dark-count rate ranges from tens to hundreds of photons per second, but this number can be drastically reduced by using coincidence counting techniques [16], up to making this effect negligible. For instance, in the implementation in [6] the dark counts where only 0.001% of the effective coincidence counts. As the dark counts occurs randomly, we cannot distinguish between a dark count and signal one. Therefore, the probability of detecting N photon in the H or V photodetector for a word a^k is finally given by

$$P_k^x(N) = \sum_{n=0}^{+\infty} \sum_{m=0}^{+\infty} \mathrm{Poi}(n; \eta^x) \mathrm{Poi}(m; \langle N_{dc} \rangle) \delta_{n+m,N} \tag{16}$$

$$= \mathrm{Poi}(N; \mu_k^x) \tag{17}$$

where $\eta^H = \langle N_c \rangle \cos^2(k\theta_m)$ and $\eta^V = \langle N_c \rangle \sin^2(k\theta_m)$, while we have defined the overall mean number of detected photons as

$$\mu_k^H = \langle N_c \rangle \cos^2(k\theta_m) + \langle N_{dc} \rangle \quad \text{and} \quad \mu_k^V = \langle N_c \rangle \sin^2(k\theta_m) + \langle N_{dc} \rangle. \tag{18}$$

As we noticed above, the dark count rate is usually very small with respect to the detected count rate of the signal. Therefore, for the H detector which detects the higher number of photons, see Equation (18), they are relevant only when $\langle N_c \rangle \sim \langle N_{dc} \rangle$. On the contrary, for the V detector, detecting the lower number of photons, their role is fundamental in determining the performance of the photonic automaton, as $\mu_m^V = \mu_{dc} = \langle N_{dc} \rangle$. This is the main difference between the two strategies: in the first, we need to distinguish between two finite mean numbers of photon $\mu_m^H = \langle N_c \rangle + \langle N_{dc} \rangle$ and $\mu_{k_1}^H$, while in the second case, we need to distinguish between the noise due to dark counts, being $\mu_m^V = \langle N_{dc} \rangle$, and $\mu_{k_1}^V$. However, to assess the performance of second strategy with respect the first one, we need to evaluate the probability of errors in the two cases.

Let us first find the threshold values in the two different strategy. We need to find the intersection between two Poissonian distributions for a word belong to L_m and a word a^{k_1} with highest probability of being erroneously being accepted, as show in Figure 3. By imposing $\mathrm{Poi}(N_{th}^x; \mu_1^x) = \mathrm{Poi}(N_{th}^x; \mu_m^x)$, where $x = H, V$, we find an exact solution for N_{th}^x given by (see the vertical dashed line in Figure 3)

$$N_{th}^x = \frac{\mu_m^x - \mu_{k_1}^x}{\ln \mu_m^x - \ln \mu_{k_1}^x}. \tag{19}$$

To highlight the dark counts effects, we introduce the ratio $\eta = \langle N_{dc} \rangle / \langle N_c \rangle$, and we have

$$N_{th}^H = \frac{\langle N_c \rangle \sin^2 \theta_m}{\ln(1 + \eta) - \ln(\cos^2 \theta_m + \eta)}, \tag{20}$$

$$N_{th}^V = \frac{\langle N_c \rangle \sin^2 \theta_m}{\ln(\sin^2 \theta_m + \eta) - \ln(\eta)}. \tag{21}$$

In our framework, the accepting problem is introduced as binary discrimination between the correct word and the word with the highest probability of error. However, in the photonic realization of the automata [6], when the number of input photons is small and m is large, also word with larger $k(\mathrm{mod}\ m)$ may contribute to the error. For this reason, like in the ideal case, we establish the minimum number of input photon $\langle N_c \rangle^{min}$ which are necessary to faithfully consider the problem as binary discrimination.

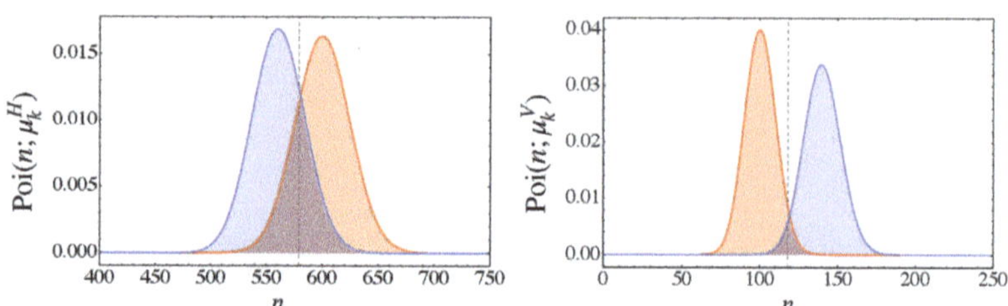

Figure 3. Probability density function of the Poissonian distribution in Equation (17) for the H detector (**left plot**) and the V detector (**right plot**) for $\langle N_c \rangle = 500$, $N_{dc} = 100$ and $m = 11$. The probability of error in Equations (25) and (29) are, respectively, $p_e^V = 0.034$ (V detector) and $p_e^H = 0.205$ (H detector). The gray dashed line is the threshold values in Equation (19). The values of the involved parameters have been chosen to better highlight the investigated effect.

To have faithfully binary discrimination the fluctuations due to the word with the second-largest probability of error, i.e., a word a^{k_2} with $k_2(\mathrm{mod}\ m) = 2$, must be much larger than the fluctuations due to the correct word, where here for "large" we mean at least two standard deviations. In this way, the discrimination can be considered only between the words a^m and a^{k_1}. In the case of a Poissonian random variable, the standard deviation is the square root of the mean for Poissonian random variables. Hence, we have the two conditions, respectively, for the H and V detector

$$\mu_{k_2}^H + 2\sqrt{\mu_{k_2}^H} < \mu_m^H - 2\sqrt{\mu_m^H}, \tag{22}$$

$$\mu_{Dc} + 2\sqrt{\mu_{Dc}} < \mu_{k_2}^V - 2\sqrt{\mu_{k_2}^V}. \tag{23}$$

In the first one we ask that the fluctuations due to the word with the second largest probability of error, i.e., a word k_2 with $k_2(\mathrm{mod}\ m) = 2$ are much larger than the fluctuations due to dark counts. In a similar way, we define the threshold for the horizontal detector. These equations can be solved for $\langle N_c \rangle$ and set a lower bounds for it such that the probability of error can be evaluated in term of a binary discrimination problem, as shown in Figure 2 (red and blue lines and points) .

Now, we can evaluate the probability of error for the two strategies. Indeed, this is equal to

$$p_e^x = p(a^{k_1})p^x(a^{k_1} \to a^m) + p(a^m)p^x(a^m \to a^{k_1}), \tag{24}$$

where we have denoted $p^x(a^i \to a^j)$ as the probability of detecting the word a^i as a^j by the detector $x = H, V$. As we have no a priori knowledge on the input word we set the prior probabilities $p(a^{k_1}) = p(a^m) = 1/2$, and for the V detector we obtain

$$P_e^V = \frac{1}{2}\left[\sum_{n=0}^{\lfloor N_{th}^V \rfloor} \frac{(\mu_{k_1}^V)^n e^{-\mu_{k_1}^V}}{n!} + \sum_{n=\lfloor N_{th}^V \rfloor + 1}^{+\infty} \frac{\mu_{dc}^n e^{-\mu_{dc}}}{n!} \right] \tag{25}$$

$$= \frac{1}{2}\left[1 - \frac{\Gamma(\lfloor N_{th}^V \rfloor + 1, \mu_{dc}) - \Gamma(\lfloor N_{th}^V \rfloor + 1, \mu_{k_1}^V)}{\lfloor N_{th}^V \rfloor!} \right] \tag{26}$$

$$= \frac{1}{2}\left[1 - \int_{N_{dc}}^{N_c \sin^2 \theta_m + N_{dc}} \frac{e^{-t} t^{\lfloor N_{th}^V \rfloor}}{\lfloor N_{th}^V \rfloor!} dt \right], \tag{27}$$

where $\Gamma(a, x)$ is the incomplete Gamma function

$$\Gamma(a, x) = \int_x^{+\infty} e^{-t} t^{a-1} dt. \tag{28}$$

Analogously, we may evaluate the probability of error for the detection of a horizontally polarized photon, i.e.,

$$P_e^H = \frac{1}{2} \left[\sum_{n=0}^{\lfloor N_{th}^H \rfloor} \frac{(\mu_m^H)^n e^{-\mu_m^H}}{n!} + \sum_{n=\lfloor N_{th}^H \rfloor + 1}^{+\infty} \frac{(\mu_{k_1}^H)^n e^{-\mu_{k_1}^H}}{n!} \right]. \tag{29}$$

Eventually, we can introduce a third strategy that combines the two described so far: for each beam of photon, we propose to measure both the H and V polarization and to combine the results so obtained. From a theoretical point of view, this is equivalent to the automata presented before, as in the ideal case the two detectors perfectly agree, i.e., one sees the photon and the other one does not see it. However, in the non-ideal case, noisy fluctuations affect photodetection. As the fluctuations in the H detector are independent from the one in the V detector, the probability of erroneously accepting it by looking at both H and V is given as

$$p_e^J = p(a^{k_1}) p^H(a^{k_1} \to a^m) p^V(a^{k_1} \to a^m) + p(a^m) p^H(a^m \to a^{k_1}) p^V(a^m \to a^{k_1}) \tag{30}$$

where J here stands for *joint*.

6. Numerical Results and Simulations

The comparison of the three strategies is reported in Figure 4. We see that the V strategy outperforms the H strategy for all the possible values of input photon, reaching almost a negligible error for approximately an order of magnitude less than the H strategy. The joint strategy realizes a further enhancement, even though the p_e^J approaches 0 with the same order of magnitude of $\langle N_c \rangle$ as p_e^V. We have also reported the solution for the inequalities (22) and (23) as a point along the corresponding line: for smaller value, the probabilities of error are not reliable as the contribution of the words with larger $k(\text{mod } m)$ is not negligible. In addition, increasing the average number of dark counts slightly increases the probability of error for all the strategies considered, even though no significant effects are detected for the considered range of values of $\langle N_{dc} \rangle$.

In Figure 5, we show the number of counts at the H and V detectors from a simulated experiment. We can see a significant reduction of the fluctuations in the V detector, which is also marked by the significant difference in the probability of error p_e^H and p_e^V. The main reason is that the counts in the V detector are affected only by the randomness due to the dark counts (if present), while in the H detector the expected number of photons contributes to the randomness of the outcomes as well. We have also reported the results for words of length $k(\text{mod } m) = 2$, which are are significantly separated from $N_c^x(m)$ and $N_c^x(k_1)$ as the value of $\langle N_c \rangle$ considered is much larger than the threshold given in (22) and (23).

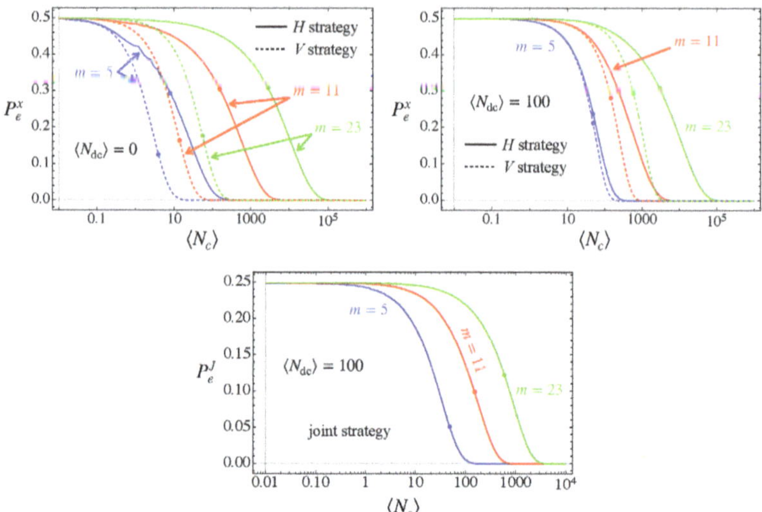

Figure 4. Probability of error for the different strategies as functions of the average number of input photon $\langle N_c \rangle$ in a semi-log plot. Red line: $m = 5$; blue line: $m = 11$; green line $m = 23$. (**Top panels**): H (solid lines) and V (dashed lines) strategies in the absence of dark counts (**left**) and for $\langle N_{dc} \rangle = 100$ (**right**). (**Bottom panel**): joint strategy in the case $\langle N_{dc} \rangle = 100$. The dots on the lines refer to the threshold values evaluated according to (22) and (23). See the text for details.

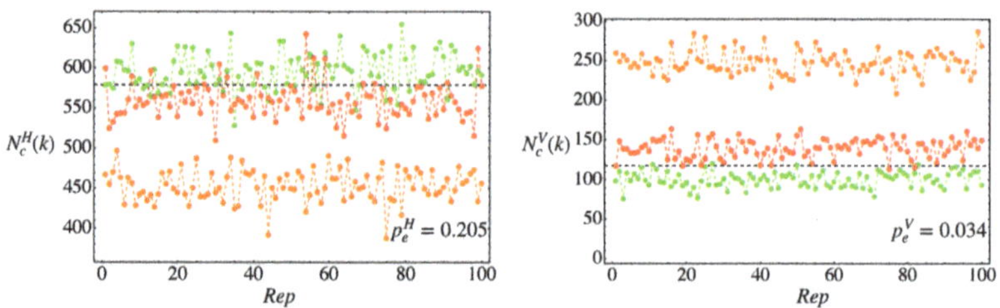

Figure 5. Simulation of $N_c^x(k)$ for the horizontal (**left**) and vertical (**right**) automata as a function of the experimental run number (*Rep*). Green dot: $k = m = 11$, i.e., $k(\mathrm{mod}\, m) = 0$; red dot: $k = 12$, i.e., $k(\mathrm{mod}\, m) = 1$; orange dot: $k = 13$, i.e., $k(\mathrm{mod}\, m) = 2$; black dashed line: N_{th}^x. We considered $\langle N_{dc} \rangle = 100$ and $\langle N_c \rangle = 500$ (the same parameters of Figure 3). The probabilities of error given in Equations (25) and (29) are respectively $p_e^V = 0.034$ and $p_e^H = 0.205$. The minimum number of input $\langle N_c \rangle$ for the H detector, solution of (22), is $\langle N_c^H \rangle^{min} = 238$, while for V, solution of (23), is $\langle N_c^V \rangle^{min} = 151$.

7. Conclusions

In this work, we have presented an enhanced photonic implementation of 1qfa for the recognition of unary language that significantly improves the performance obtained by the one originally proposed in [6]. The protocol uses the polarization degree of freedom of single photons, and exploits the possibility of detecting not only the horizontal polarization, as in [6], but also the vertical one. The resulting scheme largely outperforms the original automaton for smaller values of the mean number of sent photon $\langle N_c \rangle$. In addition, we have extended the results previously found with a detailed analysis of the conditions for which such 1qfa can work with high reliability. We have evaluated the minimum number of photons that must be sent in order to solve faithfully the inherent binary discrimination problem. As one would expect, the minimum $\langle N_c \rangle$ is smaller for the automaton that relies on the new strategy based on the V detector.

In our analysis, we have discussed the presence of dark counts in the detection of both strategies, and we have evaluated their effects both on the probability of error and on the minimum $\langle N_c \rangle$. Eventually, we also examined a joint strategy in which we combine both the H and the V detection, which can indeed be used at no additional cost. We have therefore proved that when the number of sent photon is constrained to small values, the V detection version of the 1qfa should be preferred.

Our results pave the way to the effective implementation of 1qfa using quantum optical platform, thus opening the possibility of processing strings of input symbols using feasible devices and, in turn, to introduce quantum languages and compare the complexity of classes of languages in classical and quantum cases. More generally, as the assessment of the actual power of quantum computers is one of the most significant challenges of quantum technology, implementing quantum automata provides a relevant arena to better understand the computing capabilities offered by quantum devices.

Author Contributions: Conceptualization, A.C., C.M., B.P., S.C., M.G.A.P. and S.O.; Data curation, A.C., S.C. and S.O.; Investigation, A.C., C.M., B.P., S.C., M.G.A.P. and S.O.; Methodology, A.C., C.M., B.P., S.C., M.G.A.P. and S.O.; Supervision, S.O.; Writing—original draft, A.C., C.M., B.P., S.C., M.G.A.P. and S.O.; Writing—review and editing, A.C., C.M., B.P., S.C., M.G.A.P. and S.O. All authors have read and agreed to the published version of the manuscript.

Funding: This research received no external funding.

Institutional Review Board Statement: Not applicable.

Informed Consent Statement: Not applicable.

Acknowledgments: M. G. A. Paris is member of GNFM-INdAM. C. Mereghetti and B. Palano are members of GNCS-INdAM. We thank V. Vento for useful discussion.

Conflicts of Interest: The authors declare no conflicts of interest.

Abbreviations

The following abbreviations are used in this manuscript:

1qfa	measure-once one-way quantum finite automaton
PBS	polarizing beam splitter
dc	dark counts

References

1. Castelvecchi, D. Quantum computers ready to leap out of the lab in 2017. *Nature* **2017**, *541*, 9–10. [CrossRef]
2. Ambainis, A.; Yakaryılmaz, A. Automata and Quantum Computing. *arXiv* **2018**, arXiv:cs.FL/1507.01988.
3. Bhatia, A.S.; Kumar, A. Quantum finite automata: Survey, status and research directions. *arXiv* **2019**, arXiv:cs.FL/1901.07992.
4. Bertoni, A.; Carpentieri, M. Regular Languages Accepted by Quantum Automata. *Inf. Comp.* **2001**, *165*, 174–182. [CrossRef]
5. Brodsky, A.; Pippenger, N. Characterizations of 1-Way Quantum Finite Automata. *SIAM J. Comput.* **2002**, *31*, 1456–1478. [CrossRef]
6. Mereghetti, C.; Palano, B.; Cialdi, S.; Vento, V.; Paris, M.G.A.; Olivares, S. Photonic realization of a quantum finite automaton. *Phys. Rev. Res.* **2020**, *2*, 013089. [CrossRef]
7. Bertoni, A.; Mereghetti, C.; Palano, B. Small size quantum automata recognizing some regular languages. *Theor. Comput. Sci.* **2005**, *340*, 394–407. [CrossRef]
8. Mereghetti, C.; Palano, B. Quantum automata for some multiperiodic languages. *Theor. Comput. Sci.* **2007**, *387*, 177–186. [CrossRef]
9. Ambainis, A.; Nahimovs, N. Improved constructions of quantum automata. *Theor. Comput. Sci.* **2009**, *410*, 1916–1922. [CrossRef]
10. Birkan, U.; Salehi, Ö.; Olejar, V.; Nurlu, C.; Yakaryılmaz, A. Implementing Quantum Finite Automata Algorithms on Noisy Devices. In *Computational Science—ICCS 2021*; Paszynski, M., Kranzlmüller, D., Krzhizhanovskaya, V.V., Dongarra, J.J., Sloot, P.M.A., Eds.; Springer International Publishing: Cham, Switzerland, 2021; pp. 3–16.
11. Cialdi, S.; Rossi, M.A.C.; Benedetti, C.; Vacchini, B.; Tamascelli, D.; Olivares, S.; Paris, M.G.A. All-optical quantum simulator of qubit noisy channels. *App. Phys. Lett.* **2017**, *110*, 081107. [CrossRef]

12. Olivares, S. Introduction to generation, manipulation and characterization of optical quantum states. *arXiv* **2021**, arXiv:quant-ph/2107.02519; to appear in *Phys. Lett. A*.
13. de Falco, D.; Tamascelli, D. Noise-assisted quantum transport and computation. *J. Phys. A Math. Theor.* **2013**, *46*, 225301. [CrossRef]
14. Tamascelli, D.; Zanetti, L. A quantum-walk-inspired adiabatic algorithm for solving graph isomorphism problems. *J. Phys. A Math. Theor.* **2014**, *47*, 325302. [CrossRef]
15. Rossi, M.A.C.; Benedetti, C.; Borrelli, M.; Maniscalco, S.; Paris, M.G.A. Continuous-time quantum walks on spatially correlated noisy lattices. *Phys. Rev. A* **2017**, *96*, 040301. [CrossRef]
16. Bachor, H.A.; Ralph, T.C. *A Guide to Experiments in Quantum Optics*; Wiley-VCH: Hoboken, NJ, USA, 2004.
17. Bednárová, Z.; Geffert, V.; Mereghetti, C.; Palano, B. Removing nondeterminism in constant height pushdown automata. *Inf. Comput.* **2014**, *237*, 257–267. [CrossRef]
18. Bianchi, M.P.; Mereghetti, C.; Palano, B.; Pighizzini, G. On the size of unary probabilistic and nondeterministic automata. *Fundam. Inform.* **2011**, *112*, 119–135. [CrossRef]
19. Choffrut, C.; Malcher, A.; Mereghetti, C.; Palano, B. First-order logics: Some characterizations and closure properties. *Acta Inform.* **2012**, *49*, 225–248. [CrossRef]
20. Jakobi, S.; Meckel, K.; Mereghetti, C.; Palano, B. Queue automata of constant length. In *Lecture Notes in Computer Science, Volume 8031, Proceedings of the International Workshop on Descriptional Complexity of Formal Systems, London, ON, Canada, 22–25 July 2013*; Jürgensen, H., Reis, R., Eds.; Springer: Berlin/Heidelberg, Germany, 2013; pp. 124–135.
21. Kutrib, M.; Malcher, A.; Mereghetti, C.; Palano, B. Descriptional Complexity of Iterated Uniform Finite-State Transducers. In *Lecture Notes in Computer Science, Volume 11612, Proceedings of the International Conference on Descriptional Complexity of Formal Systems, Košice, Slovakia, 17–19 July 2019*; Hospodár, M., Jirásková, G., Eds.; Springer: Berlin/Heidelberg, Germany, 2019; pp. 223–234.
22. Hopcroft, J.E.; Ullman, J.D. *Introduction to Automata Theory, Languages, and Computation*; Addison-Wesley: New York, NY, USA, 1979.
23. Bertoni, A.; Mereghetti, C.; Palano, B. Trace monoids with idempotent generators and measure-only quantum automata. *Nat. Comput.* **2010**, *9*, 383–395. [CrossRef]
24. Bianchi, M.P.; Mereghetti, C.; Palano, B. Quantum finite automata: Advances on Bertoni's ideas. *Theor. Comput. Sci.* **2017**, *664*, 39–53. [CrossRef]
25. Bianchi, M.P.; Palano, B. Behaviours of unary quantum automata. *Fundam. Inform.* **2010**, *104*, 1–15. [CrossRef]
26. Bednárová, Z.; Geffert, V.; Mereghetti, C.; Palano, B. Boolean language operations on nondeterministic automata with a pushdown of constant height. In *Lecture Notes in Computer Science, Volume 7913, Proceedings of the International Computer Science Symposium in Russia, Ekaterinburg, Russia, 25–29 June 2013*; Bulatov, A.A., Shur, A.M., Eds.; Springer: Berlin/Heidelberg, Germany, 2013; pp. 100–111.
27. Ginzburg, A. *Algebraic Theory of Automata*; Academic Press: Cambridge, MA, USA, 1968.
28. Feletti, C.; Mereghetti, C.; Palano, B. Uniform circle formation for swarms of opaque robots with lights. In *Lecture Notes in Computer Science, Volume 11201, Proceedings of the International Symposium on Stabilizing, Safety, and Security of Distributed Systems, Tokyo, Japan, 4–7 November 2018*; Izumi, T., Kuznetsov, P., Eds.; Springer: Berlin/Heidelberg, Germany, 2018; pp. 317–332.
29. Paz, A. *Introduction to Probabilistic Automata*; Academic Press: New York, NY, USA, 1971.
30. Mereghetti, C.; Pighizzini, G. Two-Way Automata Simulations and Unary Languages. *J. Autom. Lang. Comb.* **2000**, *5*, 287–300.
31. Mereghetti, C.; Palano, B.; Pighizzini, G. Note on the Succinctness of Deterministic, Nondeterministic, Probabilistic and Quantum Finite Automata. *RAIRO-Theor. Inf. Appl.* **2001**, *35*, 477–490. [CrossRef]
32. Loudon, R. *The Quantum Theory of Light*; OUP Oxford: Oxford, UK, 2000.

Article

Conditional Measurements with Silicon Photomultipliers

Giovanni Chesi [1,†], **Alessia Allevi** [2,3,†] **and Maria Bondani** [3,*,†]

1   Istituto Nazionale di Fisica Nucleare (INFN) Section of Pavia, Via Bassi 6, I-27100 Pavia, Italy; giovanni.chesi@pv.infn.it
2   Department of Science and High Technology, University of Insubria, Via Valleggio 11, I-22100 Como, Italy; alessia.allevi@uninsubria.it
3   Institute for Photonics and Nanotechnologies, CNR, Via Valleggio 11, I-22100 Como, Italy
*   Correspondence: maria.bondani@uninsubria.it; Tel.: +39-031-238-6252
†   These authors contributed equally to this work.

Abstract: Nonclassical states of light can be efficiently generated by performing conditional measurements. An experimental setup including Silicon Photomultipliers can currently be implemented for this purpose. However, these devices are affected by correlated noise, the optical cross talk in the first place. Here we explore the effects of cross talk on the conditional states by suitably expanding our existing model for conditional measurements with photon-number-resolving detectors. We assess the nonclassicality of the conditional states by evaluating the Fano factor and provide experimental evidence to support our results.

Keywords: conditional states; silicon photomultipliers; optical cross-talk; nonclassicality

Citation: Chesi, G.; Allevi, A.; Bondani, M. Conditional Measurements with Silicon Photomultipliers. *Appl. Sci.* **2021**, *11*, 4579. https://doi.org/10.3390/app11104579

Academic Editors: Robert W. Boyd and Jesús Liñares Beiras

Received: 7 April 2021
Accepted: 14 May 2021
Published: 17 May 2021

Publisher's Note: MDPI stays neutral with regard to jurisdictional claims in published maps and institutional affiliations.

1. Introduction

Given an entangled state, a conditional measurement, which is a scheme exploiting the reduction postulate [1], is a well-known option for the generation and manipulation of nonclassical and non-Gaussian states [2,3]. Remarkably, optical states have proven to be suitable for this task [3–8], especially in sight of Quantum Information protocols [9–11].

Here we focus on the detection of conditional states of light in the discrete-variable regime via photon-number-resolving (PNR) detectors. In particular, a novel class of PNR detectors, known as Silicon Photomultipiers (SiPMs), has recently experienced a remarkable technological improvement [12] and attracted attention for Quantum Optics applications [13–15]. Due to both their outstanding PNR capability and to their compactness and robustness, SiPMs may now be considered for discrete-variable Quantum-Information protocols [16]. Motivated by these points, we have recently tested a pair of SiPMs for the detection of nonclassical states of light [17,18]. Specifically, in [17] we generated a mesoscopic multi-mode twin-beam (TWB) state via type-I parametric down-conversion and post-selected one of the entangled beams by measuring the photon-number observable on the other one. We succeeded in assessing the nonclassicality of the detected conditional states.

However, as far as we know, the conditioning protocol via SiPMs on a TWB still lacks a full theoretical description. Indeed, the existing model of the effects of detection [3,4] does not include the influence of the major drawback of the SiPMs, i.e., the Optical Cross-Talk (OCT) [12,13,19,20]. The OCT is a process intrinsically connected to the very pixel structure of these devices. Being each pixel a single-photon avalanche diode, there is a chance that the avalanche triggered by a photon emits a secondary photon, which may fire a supplementary cell, resulting in a spurious count. Thus, the OCT influences the output statistics and may conceal the nonclassicality of the detected state.

Here we extend the model presented in [4] by including the effects of the OCT and provide a comparison with our experimental results. In Section 2 we define the positive operator-valued measure (POVM) describing photon counting affected by a limited quantum efficiency and by the OCT, and provide the tools needed for retrieving the statistics of the unconditioned and conditional states. Finally, we address the nonclassicality of the conditional states as sub-Poissonianity and recall the definition of the Fano factor.

In Section 3 we show our results. Firstly, we provide an analytic closed formula for the statistics of a multi-mode thermal state affected by the OCT. In a previous work of ours [21], we have already shown that in the single-mode case such a distribution is expressed in terms of the Fibonacci polynomials. In the present paper, we derive the distribution of the conditional state and consider the limit case of a TWB with an infinite number of modes. We also include the effect of the imbalance between the quantum efficiencies of the detectors. We show the effects of the OCT on the first moment of the statistics and, finally, we provide the Fano factor of the output conditioned distribution.

In Section 4 the theoretical predictions of the developed model are compared with the data from our experiment. In the same Section we also discuss how the OCT affects the light statistics and especially the consequences for the nonclassicality of the conditional states.

In Section 5 we draw our conclusions and suggest further improvements to our model.

2. Materials and Methods

2.1. Theoretical Description

We provide here all the theoretical tools needed to describe post-selection measurements in the presence of the OCT. We start with the effects of the OCT on the statistics of a multi-mode TWB, then, we derive the expression of the resulting conditional state, and finally, we show how we estimate the nonclassicality of such a state in terms of sub-Poissonianity.

2.1.1. Detected-Event Statistics of a TWB in the Presence of the OCT

A TWB is a multi-mode entangled state of light generated through a nonlinear process known as parametric down conversion [22], which is investigated in the specific context of the photon counting described in [23]. Under the assumption that the energy is equally distributed among the μ modes, a TWB state can be written as the tensor product of μ single-mode squeezed states [4,24], i.e.,

$$\hat{\Lambda} = \bigotimes_{j=1}^{\mu} |\lambda\rangle\rangle_{jj}\langle\langle\lambda| \tag{1}$$

where

$$|\lambda\rangle\rangle = \sqrt{1 - \lambda^2} \sum_n \lambda^n |n\rangle|n\rangle, \tag{2}$$

being n the number of photons, and

$$\lambda^2 \equiv \frac{N}{N + \mu} \tag{3}$$

with N as the mean number of photons in each beam. The conditioning measurement is performed on one of the two parties of the TWB state, typically named as the idler, so that the corresponding state of the other beam, which is called the signa, is ideally reduced to the same outcome, accordingly with Born's rule [1].

In the absence of the OCT effects, the POVM describing a direct measure of the photon-number operator $\hat{n}$ over multi-mode radiation reads [4]

$$\hat{\Pi}_m(\eta, \mu) = \bigotimes_{j=1}^{\mu} \sum_{l_j} \delta_{m,\gamma} \hat{\Omega}_{l_j}(\eta) \tag{4}$$

where m is the number of detected photons and $\gamma \equiv \sum_{j=1}^{\mu} l_j$, being l_j the contribution of mode j to the number of detected photons. The detection is assumed to be affected by a limited quantum efficiency η and

$$\hat{\Omega}_l(\eta) = \left(\frac{\eta}{1-\eta}\right)^l \sum_{n=l}^{\infty} \binom{n}{l}(1-\eta)^n |n\rangle\langle n| \tag{5}$$

is the single-mode photon counting POVM.

The effect of the OCT is typically described [12–14,19] by the probability ε that an avalanche triggers another single spurious avalanche from a different cell. Assuming first-order OCT events, the number of fired spurious cells cannot be larger than the number of detected photons, which implies that, for the detected event k in the presence of the OCT, we have $m \leq k \leq 2m \Rightarrow k/2 \leq m \leq k$. Note that this assumption on the OCT model is quite strong. In principle, one should consider that a primary avalanche may be related to more than one OCT event [25–27]. Indeed, it may happen that more than one of the carriers in the primary avalanche triggers a secondary one, or that a secondary avalanche triggers a tertiary one as well, and the tertiary a quaternary and so on. However, here we develop a first-order OCT model since the class of SiPMs employed in the experiment is characterized by a very low cross-talk probability. Therefore, considering higher orders would be useless. Indeed, in a previous paper of ours [14] we have shown that the cross-talk probability associated to a cascade model can be assimilated to that limited to first order as long as a larger effective value of OCT is considered.

Given this picture, we generalize the POVM in Equation (4) as follows

$$\hat{\Pi}_k(\eta, \varepsilon, \mu) = \bigotimes_{j=1}^{\mu} \sum_{l_j} \delta_{k,\gamma} \hat{\Omega}_{l_j}(\eta, \varepsilon) \tag{6}$$

with

$$\hat{\Omega}_l(\eta, \varepsilon) = \left(\frac{\varepsilon}{1-\varepsilon}\right)^l \sum_{t=\lceil l/2 \rceil}^{l} \binom{t}{l-t} \left(\frac{(1-\varepsilon)^2}{\varepsilon}\right)^t \left(\frac{\eta}{1-\eta}\right)^t \sum_{n=t}^{\infty} \binom{n}{t}(1-\eta)^n |n\rangle\langle n| \tag{7}$$

where $\lceil \cdot \rceil$ is the ceiling function. It can be shown that the operator in Equation (6) is a POVM, i.e., $\hat{\Pi}_k \geq 0$ and $\sum_k \hat{\Pi}_k = \hat{\mathbb{1}}$. Hence, one can derive the expression of the joint probability of k_s detected events on the signal and k_i on the idler as

$$P(k_s, k_i) = \mathrm{Tr}_{s,i}[\hat{\Lambda} \hat{\Pi}_{k_s} \otimes \hat{\Pi}_{k_i}] \tag{8}$$

and the marginal distributions by summing $P(k_s, k_i)$ over the corresponding variable. Note that the marginal detected-event distribution of a generic radiation field in the presence of the OCT is expressed as [14]

$$p(k) = \left(\frac{\varepsilon}{1-\varepsilon}\right)^k \sum_{m=\lceil k/2 \rceil}^{k} \binom{m}{k-m} \left(\frac{\eta(1-\varepsilon)^2}{\varepsilon(1-\eta)}\right)^m \sum_{n=m}^{\infty} \binom{n}{m}(1-\eta)^n P_n \tag{9}$$

where P_n is the photon-number distribution of the field. In Section 3 we will show the explicit form of $p(k)$ for a TWB.

We remark that our model is based on experimentally accessible quantities since the only parameter connected with the pure photon statistics, which is λ in Equation (2), can be easily expressed as a function of experimental data and parameters via

$$\lambda^2 = \frac{\langle \hat{k}_i \rangle}{\langle \hat{k}_i \rangle + \eta(1+\varepsilon)\mu} \tag{10}$$

where $\langle \hat{k}_i \rangle = (1+\varepsilon)\eta N$ is the mean value of detected events in the field including all the experimental effects.

2.1.2. Detected-Event Statistics after Post-Selection

The measurement over the idler reduces the entangled counterpart, i.e., the signal, to the corresponding outcome. The expression of the conditional state can thus be retrieved from

$$\hat{\rho}_s^{(k_i)} = \frac{1}{p(k_i)} \mathrm{Tr}_i[\hat{\Lambda} \hat{\mathbb{1}}_s \otimes \hat{\Pi}_{k_i}] \tag{11}$$

where $p(k_i)$ is the marginal distribution of detected events over the idler, according to Equation (9). Hence, the distribution of detected events for the conditional states follows as

$$p^{(k_i)}(k_s) = \mathrm{Tr}[\hat{\rho}_s^{(k_i)} \hat{\Pi}_{k_s}], \tag{12}$$

which can be read as the probability of detecting k_s events in the signal arm as long as the conditioning value is k_i. Given the distribution in Equation (12), the n-th moment comes straightforward from

$$\langle \widehat{k_s^n} \rangle^{(k_i)} = \sum_{k_s} k_s^n p^{(k_i)}(k_s). \tag{13}$$

2.1.3. Nonclassicality

Sub-Poissonianity is a well-known sufficient condition for nonclassicality [28,29]. A direct and experimentally approachable estimator of sub-Poissonianity is the ratio between the variance and the mean value of the photon-number distribution, which is known as Fano factor [29]. In particular, in Section 3 we will evaluate the Fano factor for the number of detected events, i.e.,

$$F \equiv \frac{\langle \Delta \hat{k}^2 \rangle}{\langle \hat{k} \rangle} \tag{14}$$

where $\langle \Delta \hat{k}^2 \rangle = \langle \hat{k}^2 \rangle - \langle \hat{k} \rangle^2$ is the variance of the distribution. As already shown in Refs. [14,16], in the presence of an OCT probability ε, the mean value of the detected events can be written as $\langle \hat{k} \rangle = (1+\varepsilon)\langle \hat{m} \rangle$, while the variance reads as $\langle \Delta \hat{k}^2 \rangle = (1+\varepsilon)^2 \langle \Delta \hat{m}^2 \rangle + \varepsilon(1-\varepsilon)\langle \hat{m} \rangle$. The nonclassicality condition is achieved if $F < 1$. Note that just the knowledge of the first and the second moments, provided by Equation (13), is required.

As a last remark, we point out a well-known effect of the OCT which will be crucial for our considerations on the nonclassicality: by inspecting the definition of OCT, one may infer that both the mean value and the variance of the light distribution are increased by the OCT. It can be shown that this is actually the case. However, one may also ask whether this enhancement is the same for variance and mean value, i.e., if the Fano factor remains unchanged under the effect of the OCT. The answer is no [14,16]: the OCT widens the variance with respect to the mean value and thus it heavily affects the statistics of light. This effect can be easily shown by retrieving the first and second moments of an OCT-affected distribution from Equation (9) and noting that

$$\frac{\langle \Delta \hat{k}^2 \rangle - \langle \hat{k} \rangle}{\langle \hat{k} \rangle} = (1+\varepsilon)\left[\frac{\langle \Delta \hat{m}^2 \rangle - \langle \hat{m} \rangle}{\langle \hat{m} \rangle} + \frac{2\varepsilon}{(1+\varepsilon)^2}\right] \geq \frac{\langle \Delta \hat{m}^2 \rangle - \langle \hat{m} \rangle}{\langle \hat{m} \rangle} \quad \forall \varepsilon \geq 0. \tag{15}$$

2.2. Experimental Setup and Detection Apparatus

Here we provide a description of the experiment we performed and that we will discuss in Section 4 to test our theoretical predictions.

The setup used to produce conditional states is shown in Figure 1. The fundamental and the third harmonic of a Nd:YLF laser regeneratively amplified at 500 Hz are sent to a β-barium-borate nonlinear crystal (BBO1, cut angle = 37 deg, 8-mm long) to generate the fourth harmonic (262 nm, 3.5-ps pulse duration) by sum-frequency generation. This field is used to pump parametric down conversion in a second BBO crystal (BBO2, cut angle = 46.7 deg, 6-mm long) to produce TWB states in a slightly non-collinear interaction geometry. Two twin portions are spatially and spectrally selected by means of two irises and two band-pass filters centered at 523 nm. The selected light is then delivered to a pair of PNR detectors through two multi-mode fibers having a 600-μm core diameter. As to the detectors, we employed two commercial SiPMs (mod. MPPC S13360-1350CS) operated at room temperature with an overvoltage of 3V. According to the datasheet [30], in such conditions, the detectors are endowed with a quantum efficiency of 40% at 460 nm, a moderate dark-count rate (∼140 kHz), and a low cross-talk probability (∼2%). The output of each detector is amplified by a fast inverting amplifier embedded in a computer-based Caen SP5600 Power Supply and Amplification Unit, synchronously integrated by means of a boxcar gated integrator (SR250, Stanford Research Systems) and acquired. In order to reduce as much as possible the effect of SiPMs drawbacks, the light signal was integrated over a short integration gate width (10-ns long), which roughly corresponds to the width of the peak of the output trace of the detector. Thanks to this choice, the possible contributions of dark counts and afterpulses can be neglected.

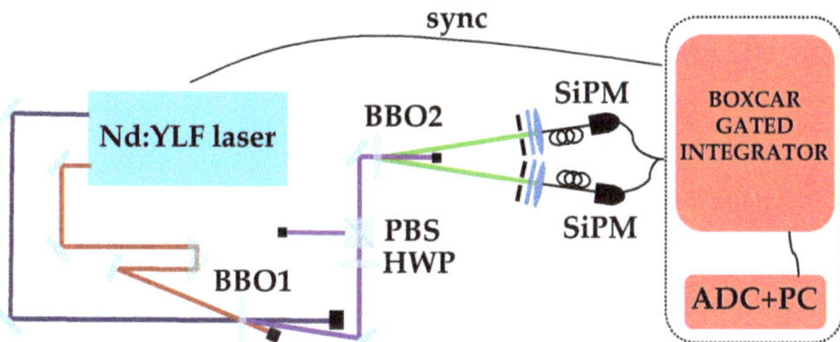

Figure 1. Setup of the experiment described in [17] and addressed in Section 4 to provide experimental evidence to the model presented here. See the text for details.

A half-wave plate (HWP) followed by a polarizing cube beam splitter (PBS) is placed on the pump beam in order to modify its intensity and thus the mean number of photons of the generated TWB states. For each mean value, 100,000 single-shot acquisitions are performed.

3. Results

3.1. The Effects of the OCT on the Photon-Number Statistics of the TWB

Here we exploit the model developed in Section 2.1 to investigate the effects of the OCT on the detection of light and, in particular, on the statistics of a multi-mode mesoscopic TWB. The topic has been already widely investigated [13,14,16,17,19]. Still, we are anyway going through this point in order to test our model and use it to provide new insights on the OCT effects implied by this description.

From the inspection of Equation (8), we find that the joint probability of detecting k_i events on the idler and k_s on the signal is given by

$$P(k_s, k_i) = (1 - \lambda^2)^\mu \left(\frac{\varepsilon}{1-\varepsilon}\right)^{k_s+k_i} \sum_{m_s=\lceil\frac{k_s}{2}\rceil}^{k_s} \sum_{m_i=\lceil\frac{k_i}{2}\rceil}^{k_i} \binom{m_s}{k_s - m_s}\binom{m_i}{k_i - m_i}$$

$$\left(\frac{(1-\varepsilon)^2}{\varepsilon}\frac{\eta}{1-\eta}\right)^{m_s+m_i} \sum_{n=\max(m_s,m_i)}^{\infty} \binom{n+\mu-1}{n}\binom{n}{m_s}\binom{n}{m_i}(\lambda(1-\eta))^{2n} \tag{16}$$

which is the extension of the joint probability retrieved in [4], where just the effect of a limited quantum efficiency is considered.

If we consider the marginal distribution from Equation (16) for the idler beam, we find

$$
\begin{aligned}
p(k_i) =& \frac{(1-\lambda^2)^\mu((1-\varepsilon)\eta\lambda^2)^{k_i}}{(1-\lambda^2(1-\eta))^{k_i+\mu}} \sum_{l=0}^{\lfloor \frac{k_i}{2} \rfloor} \binom{k_i+\mu-1-l}{\mu-1}\binom{k_i-l}{l}\left(\frac{1-\lambda^2(1-\eta)}{(1-\varepsilon)^2\eta\lambda^2}\varepsilon\right)^l \\
=& \frac{(1-\lambda^2)^\mu((1-\varepsilon)\eta\lambda^2)^{k_i}}{(1-\lambda^2(1-\eta))^{k_i+\mu}} \binom{k_i+\mu-1}{\mu-1}. \\
& {}_2F_1\left(-\frac{k_i-1}{2},-\frac{k_i}{2};-(k_i+\mu-1);-4\varepsilon\frac{1-\lambda^2(1-\eta)}{(1-\varepsilon)^2\eta\lambda^2}\right)
\end{aligned}
\tag{17}
$$

where $\lfloor \cdot \rfloor$ is the floor function and ${}_2F_1(a,b;c;x)$ is the ordinary hypergeometric function. It can be shown that Equation (17) can be obtained from Equation (9) as well by replacing P_n with the photon-number distribution of a multi-mode TWB state [6,24]. We also remark that, as we showed in [21], in the single-mode case (i.e., $\mu = 1$) Equation (17) reduces to a linear combination of Fibonacci polynomials, which should be kept in mind for the considerations that follow.

As a first remark, we stress that our model for the OCT, as outlined in Section 2.1, accounts for first-order events only. In the following paragraph, we briefly explore the implications of our simplified model for arbitrary values of ε. Then we move back to the realistic case related to our experiment.

We show the transformation of the detected-photon number statistics of TWBs due to the OCT in Figure 2 for the single-mode case and in Figure 3 for the multi-mode one. In both figures, we set the quantum efficiency $\eta = 0.17$ and the mean photon number $N = 10$ (see Equation (2)), which are experimentally reasonable values as long as SiPMs are employed for detection (see Section 4) and the photon-number regime is mesoscopic (see Section 2). For what concerns the multi-mode case, we have considered the limit $\mu \to \infty$, since, again, this case is comparable with the number of modes estimated in our experiments, where $\mu \sim 2000$ [17]. As $\mu \to \infty$, the multi-thermal distribution of TWB converges to a Poissonian one, whereas the detected-event distribution in Equation (17) tends to

$$
p_{\mu\to\infty}(k_i) = \exp\left(-\frac{\langle \hat{k}_i \rangle}{1+\varepsilon}\right)\sum_{l=0}^{\lfloor \frac{k_i}{2} \rfloor}\binom{k_i-l}{l}\frac{1}{(k_i-l)!}\left(\frac{\varepsilon}{1-\varepsilon}\right)^l\left(\frac{1-\varepsilon}{1+\varepsilon}\langle \hat{k}_i \rangle\right)^{k_i-l}.
\tag{18}
$$

Note that here we replaced the parameter λ with $\langle \hat{k}_i \rangle$ through Equation (10).

At a first glance to Figure 2, we note that the OCT gives rise to an asymmetry in the detected-event distribution: the detection probability of even events enhances proportionally to ε as the detection probability of odd events declines. Moreover, this effect is smoothed as the detected-event k_i increases. A further inspection of our OCT model may help to understand why. Let m be the number of photons detected with probability η. As mentioned above, according to our OCT model, the outcome is a number k such that $m \leq k \leq 2m$. If m is odd, then $m/2$ of the possible values for k are odd and $m/2$ are even, but, if m is even, $(m+1)/2$ of the possible values for k are even while still just $m/2$ are odd. This is basically due to the fact that $2m$, the superior bound to k, is always even. However, as m increases, such a difference between even and odd detected-photon numbers becomes negligible compared to k. This effect is especially apparent if we look at Equation (17) in the single-mode case. As mentioned above, in such a situation the detected-event distribution reduces to a linear combination of Fibonacci polynomials. This family of polynomials can be defined as [31]

$$
\mathcal{F}_n(x) \equiv \frac{1}{2^n}\frac{\left(\sqrt{x^2+4}+x\right)^n + (-1)^{n+1}\left(\sqrt{x^2+4}-x\right)^n}{\sqrt{x^2+4}}
\tag{19}
$$

for given $n \in \mathbb{N}$. The index of the polynomials in the single-mode detected-event distribution is $n = k_i + 1$, so that we get larger contributions as k_i is even and smaller otherwise.

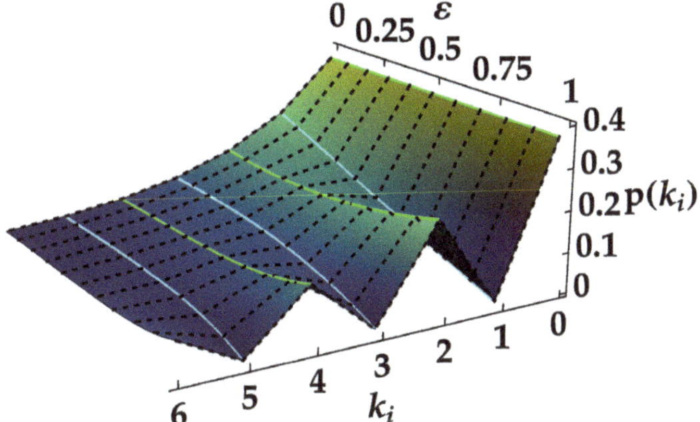

Figure 2. Detected-event distribution of the idler beam from Equation (17) in the single-mode case ($\mu = 1$) as a function of the number of detected events k_i and of the OCT probability ε. We set the quantum efficiency η to 0.17, while the mean photon number N to 10. These choices, together with a selected value of ε, yield the corresponding mean value $\langle k_i \rangle = (1 + \varepsilon)\eta N$. The plot shows the evolution of a single-thermal distribution due to the OCT.

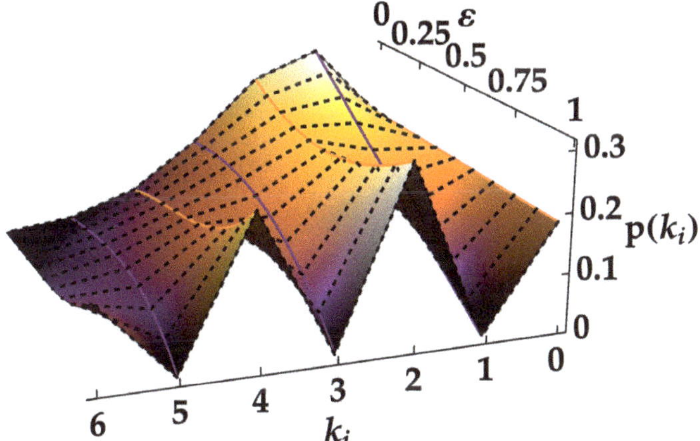

Figure 3. Detected-event distribution of the idler beam from Equation (18) in the multi-mode limit case ($\mu \to \infty$) as a function of the number of detected events k_i and of the OCT probability ε. We set the quantum efficiency η to 0.17, while the mean photon number N to 10. These choices, together with a selected value of ε, yield the corresponding mean value $\langle k_i \rangle = (1 + \varepsilon)\eta N$. The plot shows the evolution of a Poissonian distribution due to the OCT.

In Figures 4 and 5 we emphasize the most obvious effect of the OCT on the statistics of detected photons, i.e., compared to the case where no OCT affects the measurement, the probability of detecting smaller numbers of events is depleted, while, on the contrary, the larger values of k are more likely to be revealed. An expected effect of the OCT which, rather than a consequence, is the very definition of it. Note that here we focus on experimental values of ε, which are typically small ($\varepsilon < 0.1$) due to the recent technological improvements mentioned above. The plots show the ratio between the difference $\Delta \mathrm{p} \equiv \mathrm{p}(k_i) - p_0$ and p_0, where $p_0 \equiv \mathrm{p}(k_i)|_{\varepsilon=0}$. Again, we explore the single-mode case in

Figure 4, and the multi-mode limit case in Figure 5, having fixed every parameter as before. Note that the effect of the OCT in the two cases is the same, as the differences between the two plots have to be ascribed uniquely to the different distributions of pure photons, single-thermal in Figure 4 and Poissonian in Figure 5.

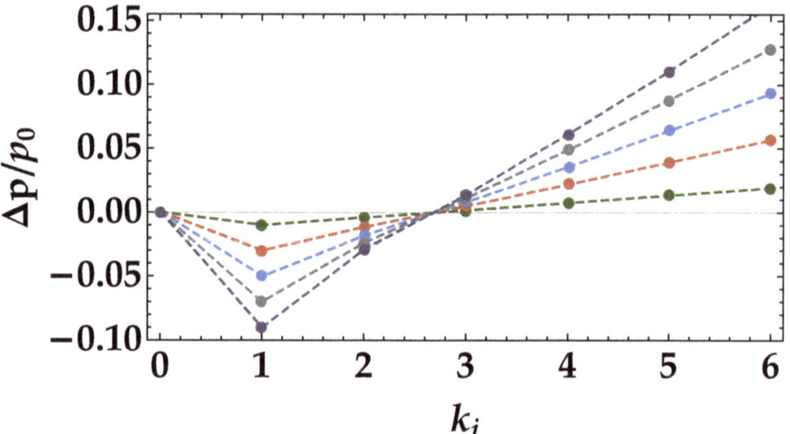

Figure 4. Plots of the relative differences $\Delta p/p_0$, with $\Delta p = p(k_i) - p_0$ and $p_0 \equiv p(k_i)|_{\varepsilon=0}$, in the single-mode case ($\mu = 1$) as a function of the number of detected events k_i, for different values of the OCT probability ε, which are $\varepsilon = 1\%$ (green), $\varepsilon = 3\%$ (red), $\varepsilon = 5\%$ (blue), $\varepsilon = 7\%$ (grey) and $\varepsilon = 9\%$ (violet). The quantum efficiency η and the mean photon number N are again set to 0.17 and 10, respectively.

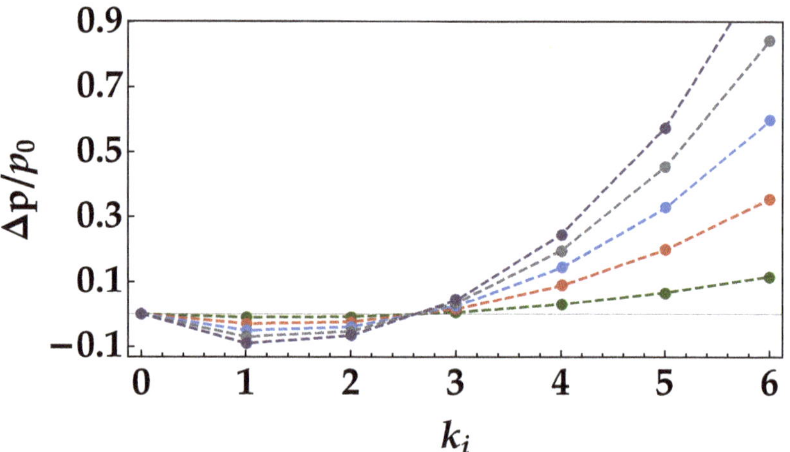

Figure 5. Plots of the relative differences $\Delta p/p_0$, with $\Delta p = p(k_i) - p_0$ and $p_0 \equiv p(k_i)|_{\varepsilon=0}$, in the multi-mode limit case ($\mu \to \infty$) as a function of the number of detected events k_i, for different values of the OCT probability ε, which are $\varepsilon = 1\%$ (green), $\varepsilon = 3\%$ (red), $\varepsilon = 5\%$ (blue), $\varepsilon = 7\%$ (grey) and $\varepsilon = 9\%$ (violet). The quantum efficiency η and the mean photon number N are again set to 0.17 and 10, respectively.

3.2. The Effects of the OCT on the Photon-Number Statistics of the Conditional State

Here we investigate the effects of the OCT on the statistics of the signal after conditioning over the idler, as described in Section 2.1. We also consider the effect of the imbalance between the quantum efficiencies of the signal and idler detectors.

Note that, while we remove the assumption that the detectors of the two parties share the same quantum efficiency $\eta = \eta_s = \eta_i$, we keep assuming the same OCT probability $\varepsilon = \varepsilon_s = \varepsilon_i$. The imbalance is introduced through the parameter $\alpha \equiv \eta_s/\eta_i$, with $\eta \equiv \eta_i$.

The expression of the reduced state of the signal after measuring k_i events over the idler is straightforward from Equation (11) and reads

$$
\rho_s^{(k_i)} = \frac{(1-\lambda^2)^\mu}{p(k_i)} \left(\frac{\varepsilon}{1-\varepsilon}\right)^{k_i} \sum_{m_i=\lceil\frac{k_i}{2}\rceil}^{k_i} \binom{m_i}{k_i - m_i} \left(\frac{(1-\varepsilon)^2}{\varepsilon}\frac{\eta}{1-\eta}\right)^{m_i}
$$
$$
\bigotimes_{j=1}^{\mu} \sum_{l_j} \delta_{m_i,\gamma} \sum_{n_j=l_j}^{\infty} \binom{n_j}{m_j} \lambda^{2n_j}(1-\eta)^{n_j} |n_j\rangle\langle n_j|
\tag{20}
$$

where again $\gamma \equiv \sum_{j=1}^{\mu} l_j$ and $\lceil\cdot\rceil$ is the ceiling function. Note that the conditional state correctly does not depend on α since no detection over the signal party has occurred yet.

On the contrary, the related detected-event distribution is a function of α, other than of the number of events detected over the idler k_i:

$$
p^{(k_i)}(k_s) = \frac{(1-\lambda^2)^\mu}{p(k_i)} \left(\frac{\varepsilon}{1-\varepsilon}\right)^{k_i+k_s} \sum_{m_i=\lceil\frac{k_i}{2}\rceil}^{k_i} \sum_{m_s=\lceil\frac{k_s}{2}\rceil}^{k_s} \binom{m_i}{k_i-m_i}\binom{m_s}{k_s-m_s}\left(\frac{(1-\varepsilon)^2}{\varepsilon}\frac{\eta}{1-\eta}\right)^{m_i}
$$
$$
\left(\frac{(1-\varepsilon)^2}{\varepsilon}\frac{\alpha\eta}{1-\alpha\eta}\right)^{m_s} \sum_{l=m_s}^{\infty} \binom{l+\mu-1}{l}\binom{l}{m_s}\binom{l}{m_i}\lambda^{2l}(1-\eta)^l(1-\alpha\eta)^l.
\tag{21}
$$

In the limit of large number of modes, we find

$$
p_{\mu\to\infty}^{(k_i)}(k_s) = \frac{\exp\left(-\frac{\langle k_i\rangle}{\eta(1+\varepsilon)}\right)}{p_{\mu\to\infty}(k_i)} \left(\frac{\varepsilon}{1-\varepsilon}\right)^{k_i+k_s} \sum_{m_i=\lceil\frac{k_i}{2}\rceil}^{k_i} \sum_{m_s=\lceil\frac{k_s}{2}\rceil}^{k_s} \binom{m_i}{k_i-m_i}\binom{m_s}{k_s-m_s}
$$
$$
\left(\frac{(1-\varepsilon)^2}{\varepsilon}\frac{\eta}{1-\eta}\right)^{m_i}\left(\frac{(1-\varepsilon)^2}{\varepsilon}\frac{\alpha\eta}{1-\alpha\eta}\right)^{m_s} \sum_{l=m_s}^{\infty} \frac{1}{l!}\binom{l}{m_s}\binom{l}{m_i}
$$
$$
\left(\frac{\langle k_i\rangle(1-\eta)(1-\alpha\eta)}{\eta(1+\varepsilon)}\right)^l.
\tag{22}
$$

Given Equation (21), we can have access to every moment of the conditional-state distribution. For instance, the first moment reads

$$
\langle \hat{k}_s\rangle^{(k_i)} = \frac{\alpha\eta(1+\varepsilon)}{1-\lambda^2(1-\eta)}\left[k_i + \mu\lambda^2(1-\eta) - \frac{\partial}{\partial x}\log\chi(x)\Big|_{x=0}\right]
\tag{23}
$$

where

$$
\chi(x) \equiv \sum_{l=0}^{\lfloor\frac{k_i}{2}\rfloor} \binom{k_i+\mu-1-l}{\mu-1}\binom{k_i-l}{l}\left(\frac{1-\lambda^2(1-\eta)}{(1-\varepsilon)^2\eta\lambda^2}\varepsilon\right)^l e^{lx}
\tag{24}
$$

is a sort of characteristic function related to the discrete probability distribution in Equation (17). Indeed, one can easily prove that Equation (17) can be rewritten as

$$
p(k_i) = \frac{(1-\lambda^2)^\mu[(1-\varepsilon)\eta\lambda^2]^{k_i}}{[1-\lambda^2(1-\eta)]^{k_i+\mu}}\chi(0).
\tag{25}
$$

The logarithmic derivatives of $\chi(x)$ evaluated in $x=0$ contribute to the moments of the conditional state, as shown in Equation (23) for the mean value. If ε is set to 0 and α to 1 in Equation (23), we retrieve the result reported in [4] for the limited-quantum-efficiency condition, i.e.,

$$\langle \hat{k}_s \rangle^{(k_i)}(\varepsilon = 0, \alpha = 1) = \frac{k_i(\langle \hat{k}_i \rangle + \eta\mu) + \mu\langle \hat{k}_i \rangle(1 - \eta)}{\langle \hat{k}_i \rangle + \mu}. \tag{26}$$

3.3. The Effects of the OCT on the Nonclassicality of the Conditional State

Finally, we focus on the nonclassicality of the state generated after post-selection and evaluate to what extent the OCT is detrimental for this quantum resource.

The first and the second moments of the conditional-state distribution allow us to retrieve the Fano factor for the detected events by means of Equation (14) expressed for the operator $\hat{k}_s$. As mentioned in Section 2.1, the Fano factor provides a sufficient condition for nonclassicality. For the distribution of the conditional state in Equation (21) we find that it reads

$$\begin{aligned}
F_s^{(k_i)} =& \frac{1 + 3\varepsilon}{1 + \varepsilon} - \alpha\eta(1 + \varepsilon) + \frac{1}{\langle \hat{k}_s \rangle^{(k_i)}} \left[\frac{\alpha\eta(1 + \varepsilon)}{1 - \lambda^2(1 - \eta)} \right]^2 \left[\lambda^2(1 - \eta)(k_i + \mu) \right] - \\
& \frac{1}{\langle \hat{k}_s \rangle^{(k_i)}} \left[\frac{\alpha\eta(1 + \varepsilon)}{1 - \lambda^2(1 - \eta)} \right]^2 \cdot \frac{\partial}{\partial x} \log\chi(x) \left[\lambda^2(1 - \eta) - \frac{\partial}{\partial x} \log\left(\frac{\partial}{\partial x} \log\chi(x) \right) \right] \bigg|_{x=0}
\end{aligned} \tag{27}$$

where $\chi(x)$ is defined in Equation (24). Again, we highlight that Equation (27) can be written as a function of experimental quantities by just replacing λ with $\langle k_i \rangle$ through Equation (10). Since the expression is quite complex, in Figure 6 we show the behavior of $F_s^{(k_i)}$ as a function of the conditioning value k_i for different choices of the other parameters: in panel (a), different mean values of the unconditioned state $\langle k_i \rangle$, in panel (b), different values of the balance parameter α, in panel (c), different choices of the cross-talk probability ε, and finally in panel (d), different number of modes of the unconditioned state μ. It is worth noting that the subPoissonianity of the Fano factor can be increased by decreasing the mean value of the unconditioned state and increasing the number of modes, and by operating on the features of the detectors, namely reducing the OCT probability and increasing the balance factor.

Again, if $\varepsilon = 0$ and $\alpha = 1$, we retrieve the known expression of the Fano factor for the conditional state in the context of multi-mode TWB states and limited quantum efficiency, as outlined in [6], i.e.,

$$F_s^{(k_i)}(\varepsilon = 0, \alpha = 1) = (1 - \eta)\left[1 + \frac{\langle \hat{k}_i \rangle(k_i + \mu)(\langle \hat{k}_i \rangle + \eta\mu)}{(\langle \hat{k}_i \rangle + \mu)[(k_i + \mu)(\langle \hat{k}_i \rangle + \eta\mu) - \eta\mu(\langle \hat{k}_i \rangle + \mu)]} \right]. \tag{28}$$

Note that Equation (27) can be significantly simplified by taking the limit to realistic values for the parameters μ and ε. As mentioned above, our experimental conditions allow us to take the limit $\mu \to \infty$, which reduces the sum in Equation (24) to

$$\chi(0) \sim \left(\langle k_i \rangle \frac{1 - \varepsilon}{1 + \varepsilon} \right)^{k_i} \sum_{l=0}^{\lfloor \frac{k_i}{2} \rfloor} \frac{1}{l!(k_i - 2l)!} \left(\frac{\varepsilon(1 + \varepsilon)}{\langle \hat{k}_i \rangle(1 - \varepsilon)^2} \right)^l, \tag{29}$$

but then, being the typical OCT probabilities of modern SiPMs of the order 10^{-2}, the largest order in the argument of the sum for a given term l is

$$\frac{(\varepsilon/\langle \hat{k}_i \rangle)^l}{l!(k_i - 2l)!}. \tag{30}$$

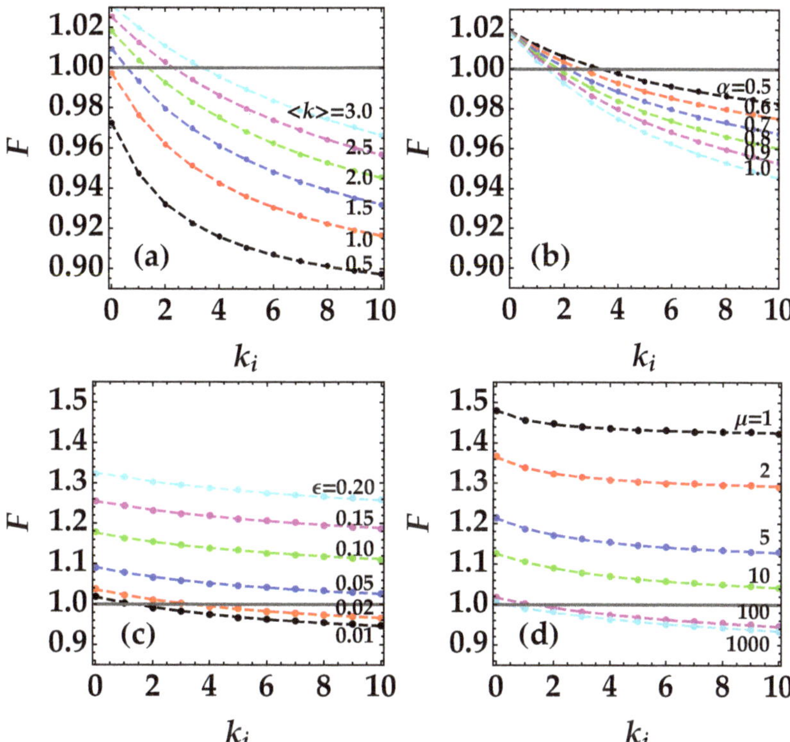

Figure 6. Fano factor of the conditional states as a function of the conditioning values for different choices of the other parameters involved in Equation (27). Panel (**a**): F for different choices of the mean value of the unconditioned state. From bottom to top: $\langle k \rangle = 0.5$ (black), $\langle k \rangle = 1$ (red), $\langle k \rangle = 1.5$ (blue), $\langle k \rangle = 2$ (green), $\langle k \rangle = 2.5$ (magenta), $\langle k \rangle = 3$ (cyan). The other parameters are: $\eta = 0.17$, $\mu = 100$, $\alpha = 1$, and $\varepsilon = 0.01$. Panel (**b**): F for different choices of the balance factor. From top to bottom: $\alpha = 0.5$ (black), $\alpha = 0.6$ (red), $\alpha = 0.7$ (blue), $\alpha = 0.8$ (green), $\alpha = 0.9$ (magenta), $\alpha = 1$ (cyan). The other parameters are: $\eta = 0.17$, $\mu = 100$, $\langle k \rangle = 2$, and $\varepsilon = 0.01$. Panel (**c**): F for different choices of the cross-talk probability. From bottom to top: $\varepsilon = 0.01$ (black), $\varepsilon = 0.02$ (red), $\varepsilon = 0.05$ (blue), $\varepsilon = 0.10$ (green), $\varepsilon = 0.15$ (magenta), $\varepsilon = 0.20$ (cyan). The other parameters are: $\eta = 0.17$, $\mu = 100$, $\langle k \rangle = 2$, and $\alpha = 1$. Panel (**d**): F for different choices of the number of modes of the unconditioned state. From top to bottom: $\mu = 1$ (black), $\mu = 2$ (red), $\mu = 5$ (blue), $\mu = 10$ (green), $\mu = 100$ (magenta), $\mu = 1000$ (cyan). The other parameters are: $\eta = 0.17$, $\langle k \rangle = 2$, $\alpha = 1$, and $\varepsilon = 0.01$.

Thus, provided that the order of the mean number of detected events is larger than the order of ε, the argument of the sum gets smaller as l increases. If we keep the $l = 0$ term only, all the logarithmic derivatives of $\chi(x)$ are null, so that the mean value and the Fano factor of the conditional state are much simplified. By taking this limit, we neglect the OCT contribution provided by the asymmetry between odd and even detected events, which is reasonable if ε is small (i.e., $\varepsilon < 0.1$), as highlighted in Figures 4 and 5. Given this approximation and the limit for μ, one gets

$$\langle \hat{k}_s \rangle^{(k_i)}_{\mu \to \infty} = \alpha[\eta(1+\varepsilon)k_i + (1-\eta)\langle \hat{k}_i \rangle]$$

$$F^{(k_i)}_{s_{\mu \to \infty}} = \frac{1+3\varepsilon}{1+\varepsilon} - \alpha\eta(1+\varepsilon)\left[1 - \frac{(1-\eta)\langle \hat{k}_i \rangle}{\eta(1+\varepsilon)k_i + (1-\eta)\langle \hat{k}_i \rangle}\right]. \tag{31}$$

Equations (31) allow us to find the threshold conditioning value $\bar{k}$ such that the detected state is nonclassical, i.e., $F^{(k_i < \bar{k})}_{s_{\mu \to \infty}} < 1$. Before retrieving $\bar{k}$, we remark that in the limit $\mu \to \infty$, the Fano factor in Equation (28), where $\varepsilon = 0$ and $\alpha = 1$, is a function of the quantum efficiency only, i.e.,

$$F^{(k_i)}_{s_{\mu \to \infty}}(\varepsilon = 0, \alpha = 1) = \frac{1 - \eta}{1 - \eta/2}. \tag{32}$$

However, $0 \leq F^{(k_i)}_{s_{\mu \to \infty}}(\varepsilon = 0, \alpha = 1) \leq 1 \, \forall \eta \in [0, 1]$, which means that the imperfections in detection due to limited quantum efficiency never provide a detected superPoissonian statistics in this context. Only in the limit case $\eta = 0$ the nonclassicality of a conditional state from multi-mode TWB is not revealed by the Fano factor, otherwise the detected nonclassicality is just reduced with respect to the ideal case ($\eta = 1$). On the contrary, the OCT can completely conceal the quantum nature of a conditional state since we may have $F^{(k_i)}_{s_{\mu \to \infty}} > 1$ for some $k_i < \bar{k}$ where

$$\bar{k} = \frac{2\varepsilon(1 - \eta)\langle \hat{k}_i \rangle}{\eta(1 + \varepsilon)[\alpha\eta(1 + \varepsilon)^2 - 2\varepsilon]}. \tag{33}$$

Note that $\bar{k} > 0 \iff \alpha\eta(1 + \varepsilon)^2 > 2\varepsilon$, i.e., if

$$\eta_s > \eta_{th}(\varepsilon) \equiv \frac{2\varepsilon}{(1 + \varepsilon)^2} \tag{34}$$

where we replaced $\alpha\eta$ with the quantum efficiency of the detector of the signal party η_s through the definition of α. Therefore, Equation (33) shows that for $\varepsilon > 0$ and $\eta < 1$ there is a conditioning number $\bar{k} > 0$ such that if $k_i < \bar{k}$ the detected statistics is superPoissonian (see Figure 6). Moreover, Equation (34) gives an experimental condition for the observation of the nonclassicality of the conditional state: provided that η_s is larger than the threshold η_{th}, then a finite $\bar{k}$ exists such that one can measure $F^{(k_i)}_{s_{\mu \to \infty}} < 1 \, \forall k_i > \bar{k}$. Note that $\bar{k}(\varepsilon = 0, \eta \neq 0) = 0$, which implies that the detected statistics is subPoissonian, if the only detection imperfection is a non-unit $\eta > 0$. However, it is remarkable that in the ideal case $\eta = 1$ we have a subPoissonian statistics independently of ε, while if $\eta \to 0$ and $\varepsilon \neq 0$, then the detected statistics is always superPoissonian, independently of k_i.

One may ask if a combination of η and ε exists such that $F^{(k_i)}_{s_{\mu \to \infty}} = 0$ for some k_i. Unfortunately, this is not the case since in the second line of Equations (31) the Fano factor is a monotone decreasing function of k_i and it converges to an asymptotic value which is strictly positive $\forall \varepsilon > 0$. Finally, we remark that the threshold in Equation (34) is directly connected to the sub-Poissonianity of the original state, which in turn depends on its intrinsic nonclassical correlations. In fact the same threshold can be shown to hold for the observation of sub-shot-noise correlations of TWB. The sub-Poissonianity condition on correlations can be expressed by the noise reduction factor as $R < 1$, where R is defined as the ratio of the variance of the difference of detected events and the mean value of their sum, i.e.,

$$R \equiv \frac{\langle \Delta(\hat{k}_s - \hat{k}_i)^2 \rangle}{\langle \hat{k}_s + \hat{k}_i \rangle}. \tag{35}$$

We showed in Ref. [20] that, in the case of TWB states with a large number of modes, this figure of merit can be reduced to

$$R = 1 - \alpha\eta(1 + \varepsilon) + \frac{2\varepsilon}{1 + \varepsilon}, \tag{36}$$

which gives $R < 1$ for the same condition as in Equation (34). Hence, the connection between the sub-Poissonianity condition and the requirement on the quantum efficiency in

Equation (34) is straightforward. Incidentally, note that the Fano factor in the second of Equations (31) can be expressed in terms of the noise reduction factor.

4. Discussion

In order to validate the model for conditioning addressed in the previous Section, hereafter we present and discuss the experimental generation of nonclassical conditional states. As already explained in [6,17,18], such states can be obtained in post-processing by selecting a certain number of photons in one TWB arm and reconstructing the modified distribution of photons in the other arm. In Section 3 we showed that the unconditioned state is formally described by a multi-thermal distribution, which reduces to Equation (18) when the light in one arm is characterized by a very large number of modes [32–34] and is detected by a SiPM characterized by an OCT probability $\varepsilon \neq 0$. In Figure 7 we show the detected-event distributions having mean values $\langle k \rangle = 2.63$ (panel (a)), 2.66 (panel (b)), 1.43 (panel (c)), and 0.57 (panel (d)). The experimental data are shown as gray dots, while the theoretical fitting functions according to Equation (18) are presented as gray lines. To quantify the agreement between the experimental data and the theoretical expectations we evaluate the fidelity $f = \sum_{m=0}^{\bar{m}} \sqrt{P^{\text{th}}(k)P(k)}$, in which $P^{\text{th}}(k)$ and $P(k)$ are the theoretical and experimental distributions, respectively, and the sum extends up to the maximum number of detected events k above which both $P^{\text{th}}(k)$ and $P(k)$ become negligible.

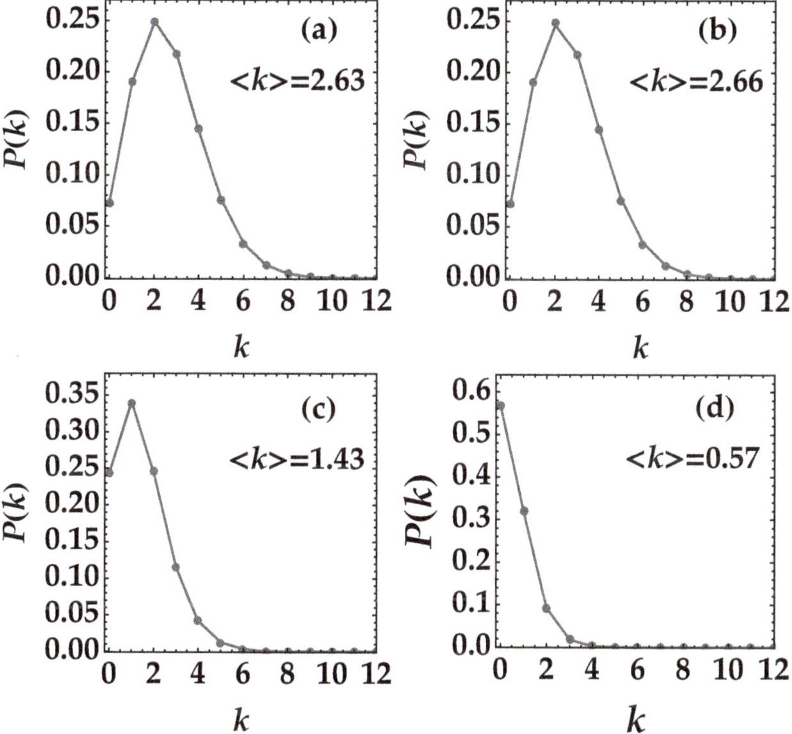

Figure 7. Detected-event distributions $P(k)$ of the unconditioned state having mean value $\langle k \rangle = 2.63$ (panel (**a**)), 2.66 (panel (**b**)), 1.43 (panel (**c**)), and 0.57 (panel (**d**)). The experimental data are shown as gray dots, while the theoretical expectations are presented as gray lines. The fidelity values are: $f = 0.9999$ in all panels.

From the fitting procedure, it is possible to obtain the value of the only fitting parameter, namely the OCT. In particular, we notice that for the four considered measurements, the OCT value is of the same order of magnitude and always less than 1%, thus proving that the cross-talk probability affecting this model of SiPM is really small, even if not completely negligible. We remark that the estimated values for the OCT probability in Figure 8 are smaller than those reported in the datasheet of our sensors [30], but consistent with the characterization that we have already provided for these SiPMs in [16].

In order to prove that the conditioning procedure changes the statistical properties of such states making them sub-Poissonian, we calculate the Fano factor of the conditional states obtained from each of the four considered unconditioned states. Indeed, as mentioned in Section 2.1, $F < 1$ is a sufficient condition for nonclassicality.

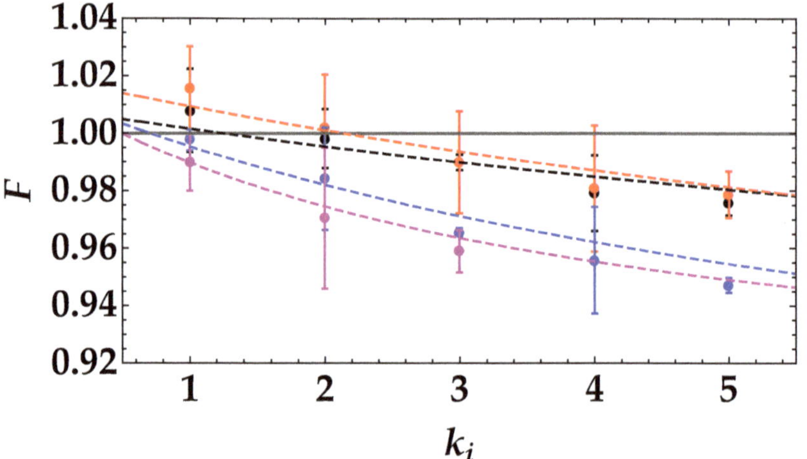

Figure 8. Fano factor as a function of the conditioning value for four different unconditioned states having mean values $\langle k \rangle = 2.63$ (black), 2.66 (red), 1.43 (blue), and 0.57 (magenta). The experimental data are shown as dots plus error bars, while the theoretical fitting functions according to the second line of Equations (31) are presented as dashed curves with the same color choice. The fitting parameters are the following: $\eta = 0.134$, $\alpha = 0.990$ (black curve), $\eta = 0.157$, $\alpha = 0.989$ (red curve), $\eta = 0.158$, $\alpha = 0.997$ (blue curve), and $\eta = 0.125$, $\alpha = 0.986$ (magenta curve). The reduced $\chi^{(2)}$ are: 0.34 (black curve), 0.14 (red curve), 0.94 (blue curve), and 0.05 (magenta curve).

In Figure 8 we show the experimental Fano factors shown as dots plus error bars, while the theoretical fitting functions according to the second line of Equations (31) are shown as dashed lines with the same color choice. For all the fitting functions we left η and α as free fitting parameters, while we used the same values of ε obtained from the fitting of the marginal distributions. In particular, in all cases we obtained a balance factor $\alpha \sim 0.99$ and a quantum efficiency $\eta \sim 0.14$. As a general statement, we note that the data corresponding to the conditioning value $k_i = 1$ are larger than 1 for the largest mean values. Such a behavior is in agreement with the theoretical expectation expressed by the second line of Equations (31) and the plots in panel (a) of Figure 6. Moreover, we emphasize that for the smallest mean value the conditioning operation is applied up to $k_i = 3$ because the number of experimental data is not sufficient to reliably build the states corresponding to $k_i > 3$.

In order to explore in which way the conditional measurements modify the statistical properties of the unconditioned states in the presence of the OCT, in the two panels of Figure 9 we show some conditional distributions at different conditioning values together with the corresponding unconditioned statistics having mean values $\langle k \rangle = 2.66$ (panel (a)), and 1.43 (panel (b)). The data are presented as colored dots plus error bars ($k_i = 1$ in black, $k_i = 2$ in red, and $k_i = 3$ in blue), while the theoretical expectations are shown as solid lines with the same color choice. The theoretical curves have been calculated according to Equation (22) using the parameter values of ε obtained from the fit of the unconditioned states (see caption of Figure 7) and those of η and α obtained from the fit of the Fano (see caption of Figure 8). For the sake of clarity, in each panel of Figure 9 we show again the statistics of the unconditioned state as gray dots and the theoretical expectation as gray surface defined by dashed line. As expected from the two panels of the figure, it clearly appears that the conditional measurements change the statistics of the input state. Even in this case, to quantify the agreement between the experimental data and the theoretical expectations we evaluate the fidelity. We note that the higher the conditioning value the lower the fidelity value. This fact can be ascribed to the limited number of data at our disposal to build the statistics, which is lower and lower at increasing values of k_i. Larger acquisitions of data could overcome such a limit. At the same time, the good dynamic range of SiPMs would suggest that both the unconditioned states and the corresponding conditional ones could be more populated, thus allowing us to really explore the mesoscopic intensity domain.

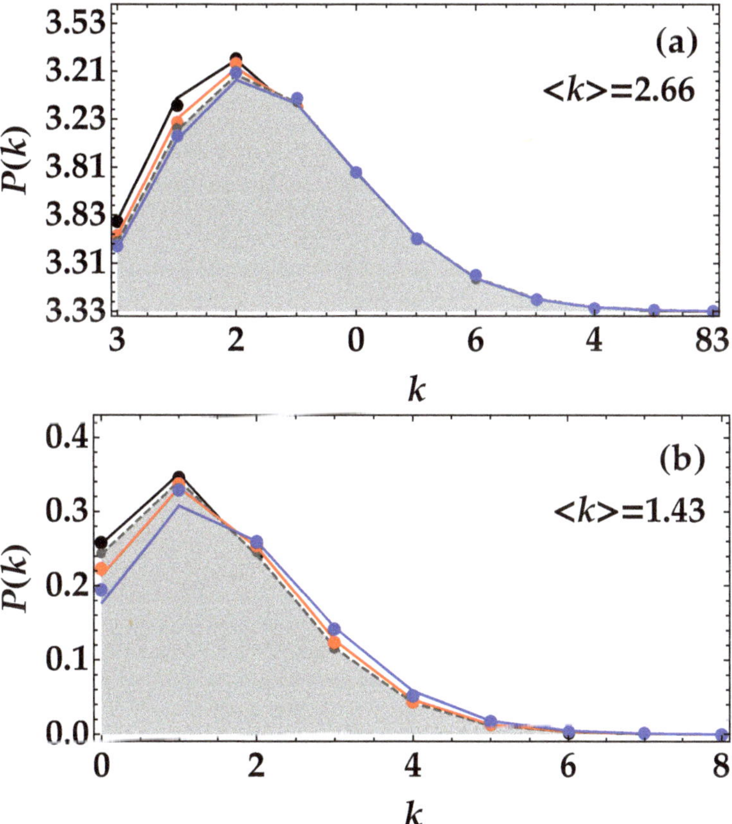

Figure 9. Detected-event distributions $P(k)$ of the conditional states for $k_i = 1$ (black curve), $k_i = 2$ (red curve), and $k_i = 3$ (blue curve) obtained from an unconditioned state having mean value $\langle k \rangle = 2.66$ (panel (**a**)) and 1.43 (panel (**b**)). The experimental data are shown as colored symbols, while the theoretical expectations are presented as solid lines with the same color choice. The fidelity values are: $f = 0.9999$, 0.9961, and 0.9888 in panel (**a**) and $f = 0.9999$, 0.9954, and 0.9863 in panel (**b**). The unconditioned state is shown as gray dots and its theoretical expectation as gray surface defined by dashed line. The fidelity value is $f = 0.9996$ in panel (**a**) and $f = 0.9999$ in panel (**b**).

In general, the good agreement between the experimental data and the theoretical expectations validate the model used to describe the role played by the non-idealities of the employed detectors, namely the cross-talk effect, the non-unitary quantum efficiency and the possible imbalance between the two quantum efficiencies. We emphasize that the detection of subPoissonian states was achieved because, even in the presence of a limited quantum efficiency, the OCT probability is small enough to ensure that $\eta > \eta_{th}$. This is not the case of either previous sensors generations, in which the OCT probability was more than 10%, or new kinds of SiPMs with a higher sensitivity in the near infrared region. Indeed, the best generation of such detectors exhibits a cross talk probability of 6% and a low quantum efficiency (less than 20%), which could prevent the generation of nonclassical states by conditional measurements.

5. Conclusions

In this paper we addressed a thorough theoretical model for the conditional measurements with SiPMs. In particular, we included the contribution of the OCT and we took into account the possibility of an imbalance between the two detection chains. We provided a complete description of the detection of a multi-mode TWB state in the presence of the OCT, showing explicitly the effects of such correlated noise on the reconstructed distribution. We obtained a closed formula for the detected-event distribution of the conditional states and an analytic expression for the first moments. Hence, we retrieved the Fano factor, which represents a sufficient criterion for nonclassicality. In particular, we found that, in the presence of cross-talk effect, nonclassicality is more easily attained by:

- reducing the imbalance between the two detection arms;
- decreasing the mean value of the unconditioned states;
- increasing the number of modes.

Moreover, we found a useful bound between the quantum efficiency of the detectors and the OCT probability, which sets a link between their mutual values for still revealing the nonclassicality of conditional states. Actually, we demonstrated that this bound is valid in general for the twin-beam states, on which the conditioning operation is performed.

The theoretical expectations have been validated by the experimental generation of conditional states by conditional measurements performed on multi-mode TWB states with SiPMs. The good agreement between the experimental data and the theoretical predictions suggests that the conditional measurements can be performed even on more populated states to produce well-populated conditional states as well by exploiting the good dynamic range of SiPMs.

Finally, we hint that the model may be further improved by including the dark counts, which is another common drawback of SiPMs at room temperature. However, it is worth noting that in the case of a light signal integrated over short gate widths [17], which correspond to our experimental condition, the mean number of dark counts is remarkably low, which is the reason why we did not address this topic here. Moreover, it could be interesting to include cascade effects and generation of multiple secondary avalanches in the model for the OCT, so that a realistic case for large ε could be compared with the one explored here.

Author Contributions: G.C., A.A. and M.B. conceptualized the work, G.C. performed the theoretical calculations, G.C. and A.A. performed the measurements, A.A. and M.B. analysed and interpreted the data, G.C. and A.A. drafted the work, M.B. substantively revised it. All authors have read and agreed to the published version of the manuscript.

Funding: This research received no external funding.

Institutional Review Board Statement: Not applicable.

Informed Consent Statement: Not applicable.

Data Availability Statement: The datasets used and analysed during the current study are available from the corresponding author on reasonable request.

Conflicts of Interest: The authors declare no conflict of interest.

Abbreviations

The following abbreviations are used in this manuscript:

PNR	Photon Number Resolving
SiPM	Silicon Photomultiplier
OCT	Optical Cross-Talk
TWB	Twin Beam
POVM	Positive Operator-Valued Measure

References

1. Paris, M.G.A. The modern tools of quantum mechanics. *Eur. Phys. J. Spec. Top.* **2012**, *203*, 61–86. [CrossRef]
2. Ben-Zion, D.; McGreevy, J.; Grover, T. Disentangling quantum matter with measurements. *Phys. Rev. B* **2020**, *101*, 115131. [CrossRef]
3. Paris, M.G.A.; Cola, M.; Bonifacio, R. Quantum-state engineering assisted by entanglement. *Phys. Rev. A* **2003**, *67*, 042104. [CrossRef]
4. Allevi, A.; Andreoni, A.; Beduini, F.A.; Bondani, M.; Genoni, M.G.; Olivares, S.; Paris, M.G.A. Conditional measurements on multimode pairwise entangled states from spontaneous parametric downconversion. *EPL* **2010**, *92*, 20007. [CrossRef]
5. Allevi, A.; Andreoni, A.; Bondani, M.; Genoni, M.G.; Olivares, S. Reliable source of conditional states from single-mode pulsed thermal fields by multiple-photon subtraction. *Phys. Rev. A* **2010**, *82*, 013816. [CrossRef]
6. Lamperti, M.; Allevi, A.; Bondani, M.; Machulka, R.; Michálek, V.; Haderka, O.; Peřina, J., Jr. Optimal sub-Poissonian light generation from twin beams by photon-number resolving detectors. *J. Opt. Soc. Am. B* **2014**, *31*, 20–25. [CrossRef]
7. Peřina, J., Jr.; Haderka, O.; Michálek, V. Simultaneous observation of higher-order non-classicalities based on experimental photocount moments and probabilities. *Sci. Rep.* **2019**, *9*, 1–8. [CrossRef]
8. Dakna, M.; Anhut, T.; Opatrný, T.; Knöll, L.; Welsch, D.G. Generating Schrödinger-cat-like states by means of conditional measurements on a beam splitter. *Phys. Rev. A* **1997**, *55*, 3184. [CrossRef]
9. Olivares, S.; Paris, M.G.A.; Bonifacio, R. Teleportation improvement by inconclusive photon subtraction. *Phys. Rev. A* **2003**, *67*, 032314. [CrossRef]
10. Cerf, N.J.; Krüger, O.; Navez, P.; Werner, R.F.; Wolf, M.M. Non-Gaussian Cloning of Quantum Coherent States is Optimal. *Phys. Rev. Let.* **2005**, *95*, 070501. [CrossRef]
11. Eisert, J.; Scheel, S.; Plenio, M.B. Distilling Gaussian States with Gaussian Operations is Impossible. *Phys. Rev. Lett.* **2002**, *89*, 137903. [CrossRef]
12. Caccia, M.; Chmill, V.; Ebolese, A.; Locatelli, M.; Martemiyanov, A.; Pieracci, M.; Risigo, F.; Santoro, R.; Tintori, C. An Educational Kit Based on a Modular Silicon Photomultiplier System. In Proceedings of the 2013 3rd International Conference on Advancements in Nuclear Instrumentation, Measurement Methods and Their Applications (ANIMMA), Marseille, France, 23–27 June 2013; Volume 978, pp. 1–7.
13. Afek, I.; Natan, A.; Ambar, O.; Silberberg, Y. Quantum state measurements using multipixel photon detectors. *Phys. Rev. A* **2009**, *79*, 043830. [CrossRef]
14. Ramilli, M.; Allevi, A.; Chmill, V.; Bondani, M.; Caccia, M.; Andreoni, A. Photon-number statistics with silicon photomultipliers. *J. Opt. Soc. Am. B* **2010**, *27*, 852–862. [CrossRef]
15. Kalashnikov, D.A.; Tan, S.H.; Iskhakov, T.S.; Chekhova, M.V.; Krivitsky, L.A. Measurement of two-mode squeezing with photon number resolving multipixel detectors. *Opt. Lett.* **2012**, *37*, 2829–2831. [CrossRef] [PubMed]
16. Chesi, G.; Malinverno, L.; Allevi, A.; Santoro, R.; Caccia, M.; Martemiyanov, A.; Bondani, M. Optimizing Silicon Photomultipliers for Quantum Optics. *Sci. Rep.* **2019**, *9*, 1–12. [CrossRef] [PubMed]
17. Chesi, G.; Malinverno, L.; Allevi, A.; Santoro, R.; Caccia, M.; Bondani, M. Measuring nonclassicality with silicon photomultipliers. *Opt. Lett.* **2019**, *44*, 1371–1374. [CrossRef] [PubMed]
18. Chesi, G.; Allevi, A.; Bondani, M. Autocorrelation functions: A useful tool for both state and detector characterisation. *Quantum Meas. Quantum Metrol.* **2019**, *6*, 1–6. [CrossRef]
19. Nagy, F.; Mazzillo, M.; Renna, L.; Valvo, G.; Sanfilippo, D.; Carbone, B.; Piana, A.; Fallica, G.; Molnar, J. Afterpulse and delayed crosstalk analysis on a STMicroelectronics silicon photomultiplier. *Nucl. Instrum. Methods Phys. Res. A* **2014**, *759*, 44–49. [CrossRef]
20. Chesi, G.; Allevi, A.; Bondani, M. Effects of nonideal features of silicon photomultiplier on the measurements of quantum correlations. *Int. J. Quantum Inf.* **2019**, *17*, 1941012. [CrossRef]
21. Chesi, G.; Allevi, A.; Bondani, M. Effect of cross-talk on conditional measurements performed with multi-pixel photon counters. In Proceedings of the 2020 IMEKO TC-4 International Symposium, Palermo, Italy, 14–16 September 2020.
22. Couteau, C. Spontaneous parametric down-conversion. *Contemp. Phys.* **2018**, *59*, 291–304. [CrossRef]

23. Paleari, F.; Andreoni, A.; Zambra, G.; Bondani, M. Thermal photon statistics in spontaneous parametric downconversion. *Opt. Lett.* **2004**, *12*, 2816–2824. [CrossRef]
24. Allevi, A.; Olivares, S.; Bondani, M. Measuring high-order photon number correlations in experiments with multimode pulsed quantum states. *Phys. Rev. A* **2012**, *85*, 063835. [CrossRef]
25. Van Dam, H.T.; Seifert, S.; Vinke, R.; Dendooven, P.; Löhner, H.; Beekman, F.J.; Schaart, D.R. A Comprehensive Model of the Response of Silicon Photomultipliers. *IEEE Trans. Nucl. Sci.* **2010**, *57*, 2254–2266. [CrossRef]
26. Vinogradov, S. Analytical models of probability distribution and excess noise factor of solid sate photomultiplier signals with crosstalk. *Nucl. Instruments Methods Phys. Res. Sect. A Accel. Spectrometers Detect. Assoc. Equipment* **2012**, *695*, 247–251. [CrossRef]
27. Gallego, L.; Rosado, J.; Blanco, F.; Arqueros, F. Modeling crosstalk in silicon photomultipliers. *J. Instrum.* **2013**, *8*, P05010. [CrossRef]
28. Glauber, R.J. The Quantum Theory of Optical Coherence. *Phys. Rev.* **1963**, *130*, 2529–2539. [CrossRef]
29. Klyshko, D.N. The nonclassical light. *Phys.-Usp.* **1996**, *39*, 573–596. [CrossRef]
30. MPPC S13360-1350CS. Available online: http://www.hamamatsu.com/us/en/S13360-1350CS.html (accessed on 17 May 2021).
31. Bicknell, M. A primer for the Fibonacci numbers VII. *Fibonacci Quart.* **1970**, *8*, 407–420.
32. Allevi, A.; Lamperti, M.; Bondani, M.; Peřina, J., Jr.; Michálek, V.; Haderka, O.; Machulka, R. Characterizing the nonclassicality of mesoscopic optical twin-beam states. *Phys. Rev. A* **2013**, *88*, 063807. [CrossRef]
33. Machulka, R.; Haderka, O.; Peřina, J., Jr.; Lamperti, M.; Allevi, A.; Bondani, M. Spatial properties of twin-beam correlations at low- to high-intensity transition. *Opt. Express* **2014**, *22*, 13374–13379. [CrossRef] [PubMed]
34. Allevi, A.; Bondani, M. Statistics of twin-beam states by photon-number resolving detectors up to pump depletion. *J. Opt. Soc. Am. B* **2014**, *31*, B14–B19. [CrossRef]

applied sciences

Article

Experimental Quantum Message Authentication with Single Qubit Unitary Operation

Min-Sung Kang [1,2], Yong-Su Kim [1,3], Ji-Woong Choi [1,4], Hyung-Jin Yang [4] and Sang-Wook Han [1,3,*]

[1] Center for Quantum Information, Korea Institute of Science and Technology (KIST), Seoul 02792, Korea; mskang81@korea.kr (M.-S.K.); yong-su.kim@kist.re.kr (Y.-S.K.); jodol007@kist.re.kr (J.-W.C.)
[2] Korean Intellectual Property Office (KIPO), Government Complex Daejeon Building 4, 189, Cheongsa-ro, Seo-gu, Daejeon 35208, Korea
[3] Division of Nano and Information Technology, Korea Institute of Science and Technology School, Korea University of Science and Technology, Seoul 02792, Korea
[4] Department of Physics, Korea University, Sejong 30019, Korea; yangh@korea.ac.kr
[*] Correspondence: swhan@kist.re.kr; Tel.: +82-31-546-7474

Abstract: We have developed a quantum message authentication protocol that provides authentication and integrity of an original message using single qubit unitary operations. Our protocol mainly consists of two parts: quantum encryption and a correspondence check. The quantum encryption part is implemented using linear combinations of wave plates, and the correspondence check is performed using Hong–Ou–Mandel interference. By analyzing the coincidence counts of the Hong–Ou–Mandel interference, we have successfully proven the proposed protocol experimentally, and also showed its robustness against an existential forgery.

Keywords: quantum message authentication; quantum three-pass protocol; Gao's forgery; swap test

Citation: Kang, M.-S.; Kim, Y.-S.; Choi, J.-W.; Yang, H.-J.; Han, S.-W. Experimental Quantum Message Authentication with Single Qubit Unitary Operation. *Appl. Sci.* **2021**, *11*, 2653. https://doi.org/10.3390/app11062653

Academic Editors: Maria Bondani, Alessia Allevi and Stefano Olivares

Received: 28 February 2021
Accepted: 13 March 2021
Published: 16 March 2021

Publisher's Note: MDPI stays neutral with regard to jurisdictional claims in published maps and institutional affiliations.

1. Introduction

Modern cryptography provides four functions, namely, confidentiality, authentication, integrity, and nonrepudiation [1,2]. Therefore, as a substitution candidate for next-level secure cryptography, quantum cryptography should also have the ability to offer these four functions. Remarkable progress has been made in the area of confidentiality because the quantum key distribution (QKD) protocol that provides confidentiality has been considerably improved [3–6]. QKD aims to enable communication partners, e.g., Alice and Bob, to share secret keys and ultimately perform a one-time pad communication. Those protocols provide unconditional confidentiality based on the principle that an arbitrary unknown quantum state cannot be copied and that quantum measurement is irreversible [7–10]. On the other hand, many researchers have also studied how to use these secret keys in quantum message authentication [11–13], arbitrated quantum signature [14–19], or quantum digital signature [20–29], providing authentication, integrity, and non-repudiation.

In this paper, we introduce a simple and practical quantum message authentication protocol with a quantum three-pass protocol [30–33] and a quantum encryption scheme [19,34]. This protocol is a lightweight to simplify the implementation by removing an arbitrator from our proposed quantum signature protocol [19]. Here, the quantum three-pass protocol is the quantum version of Shamir's three-pass protocol [1,35], and quantum encryption scheme is to prevent existential forgery, called Gao's forgery. More specifically, the core elements of the proposed protocol, such as the quantum three-pass protocol and the quantum encryption scheme, are implemented with only single qubit unitary operators. In other words, these can be implemented easily by using linear combinations of wave plates [36,37]. Additionally, the swap test that checks the correspondence of the original message and quantum message authentication code (QMAC) can be implemented using a Hong–Ou–Mandel interferometer [38–40]. In advance, as the Hong-Ou-Mandel interferometer is a destructive swap test [40], more resources are needed to implement a controlled swap test.

In Section 2, we briefly explain the concept of the proposed scheme. Section 3 presents a security analysis of the proposed protocol for Alice's private key, the forgery of QMAC pair, and the origin authentication of quantum message. Section 4 describes the experimental setup and measurement results. We conducted three experiments with the proposed protocol. First, we implemented a quantum three pass protocol, which is a method of conveying information in the proposed quantum message authentication. Second, we implemented a quantum encryption scheme with a single qubit unitary operator to prevent forgery. Finally, we confirmed that the QMAC pair with the quantum encryption scheme is robust to Gao's forgery. In Section 5, after a thorough discussion that includes the possibility of expanding the scheme to quantum signature and quantum entity authentication, we present the conclusions of this work.

2. Quantum Message Authentication Protocol

Quantum message authentication, which is similar to conventional message authentication, should provide message integrity and origin authentication. What differentiates quantum message authentication from conventional message authentication [41,42] is that the former uses quantum states $|0\rangle$ and $|1\rangle$ as a message represented by a sequence of "0" and "1" bits. In addition, using arbitrary quantum states as a message enables more information to be delivered at once [43,44]. Moreover, there is a significant difference that is described below. In modern cryptography, asymmetric key cryptography easily provides message integrity, message origin authentication, and nonrepudiation. Unfortunately, a quantum asymmetric key cryptosystem based on the quantum trapdoor one-way function do not exist, making the design of quantum authentication and quantum signature protocols difficult. To overcome this difficulty, we propose a new quantum message authentication protocol based on Shamir's three-pass protocol [1,35]. Shamir's three pass protocol has the advantage that two parties, e.g., Alice and Bob, can share information without exposing their own private keys. In the implementation, the central idea is that the commutative property [19] of exponential operation in Shamir's three-pass protocol is implemented using single-qubit rotation operators consisting of linear combinations of wave plates. To our knowledge, this is the first time a quantum message authentication protocol has been proposed using the quantum three-pass protocol, though other applications of the quantum three-pass protocol, such as direct communication [32], quantum key distribution [30], and quantum signature [19], have been proposed theoretically. Figure 1 schematically shows the quantum message authentication protocol that we implemented. Our quantum message authentication protocol consists of preparation, quantum message authentication, and verification phase.

2.1. Preparation Phase

In the preparation phase, Alice and Bob pre-share secret key sequences $K_{AB} = \left(k_{AB}^1, k_{AB}^2, \ldots, k_{AB}^N \right)$ and $K_H = \left(k_H^1, k_H^2, \ldots, k_H^N \right)$ that determine which single-qubit operation is chosen. The sequences $K_{AB} = \left(k_{AB}^1, k_{AB}^2, \ldots, k_{AB}^N \right)$ and $K_H = \left(k_H^1, k_H^2, \ldots, k_H^N \right)$ are a classical bit sequence with the size of $2N$ and N respectively, where $k_{AB}^i \in \{00, 01, 10, 11\}$, $k_H^i \in \{0, 1\}$. The secret keys k_{AB}^i and k_H^i correspond to the Pauli operators $\sigma_{k_{AB}^i} \in \{I, \sigma_x, \sigma_y, \sigma_z\}$ and the operator $H^{k_H^i} \in \{H^0 = I, H^1 = H\}$. Here, operator is a linear combination of the Pauli operators $\{I, \sigma_x, \sigma_y, \sigma_z\}$ and unitary operator $H^\dagger H = HH^\dagger = I$.

$$H = \left(I - i\sigma_x - i\sigma_y - i\sigma_z \right)/2 \tag{1}$$

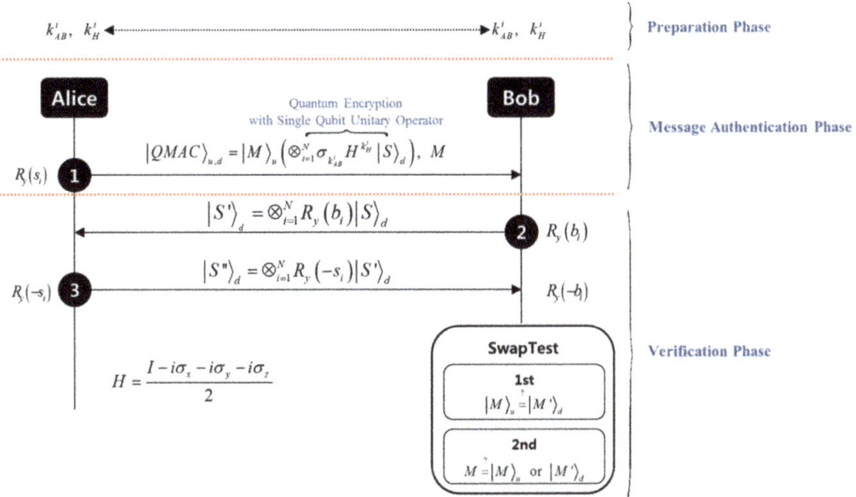

Figure 1. Basic structure of the quantum message authentication protocol based on quantum three-pass protocol. Similar to quantum three-pass protocol, which transmits bits three times, our protocol performs three quantum state transmissions. After three attempts of quantum state transmission, Bob finally acquires quantum message states $|M\rangle_u = \otimes_{i=1}^{N} R_y(m_i)|\varphi\rangle_u^i$ and $|M'\rangle_d = \otimes_{i=1}^{N} R_y(m_i)|\varphi\rangle_d^i$. He then uses a swap test twice to confirm the similarity of the two arbitrary quantum states $|M\rangle_u, |M'\rangle_d$ and bit message sequence M. K_{AB} and K_H denote the secret key sequences that Alice and Bob previously shared. S is the private key sequence that only Alice knows, and B is the one known only to Bob.

2.2. Quantum Message Authentication Phase

The quantum message authentication phase is composed of two stages: quantum message generation, QMAC generation, and quantum encryption. In the quantum message generation stage, Alice generates a quantum message state pair

$$|M\rangle_u|M\rangle_d = \left[\otimes_{i=1}^{N} R_y(m_i)|\varphi\rangle_u^{(i)}\right]\left[\otimes_{i=1}^{N} R_y(m_i)|\varphi\rangle_d^{(i)}\right] \tag{2}$$

by applying a single qubit rotation operator

$$R_y(m_i) = \begin{pmatrix} \cos\frac{m_i}{2} & -\sin\frac{m_i}{2} \\ \sin\frac{m_i}{2} & \cos\frac{m_i}{2} \end{pmatrix}, \tag{3}$$

where $M = (m_1, m_2, m_3, \ldots, m_N)$ is a rotation angle sequence, $0° \leq m_i \leq 360°$, and $|\varphi\rangle_u^{(i)}|\varphi\rangle_d^{(i)}$ are the logical states $|0\rangle|0\rangle$ or $|1\rangle|1\rangle$, corresponding to horizontally polarized photons $|H\rangle|H\rangle$ and vertically polarized photons $|V\rangle|V\rangle$, respectively. The superscript (i) denotes the i th qubit, and subscripts u and d denote up and down, corresponding to the up-line and down-line, respectively, of the experimental setup used for our protocol. The rotation angle sequence $M = (m_1, m_2, m_3, \ldots, m_N)$ is a bit message sequence, and we assume that it has already been published in public as in the case of a contract or an official document. The reason for publishing M is to prevent Alice from attempting to forge using a modulated QMAC pair, which is discussed in detail in Section 3.2 impossibility of forgery.

In the QMAC generation stage, Alice encrypts the quantum message pair $|M\rangle_u|M\rangle_d$ of Equation (2) by using a single qubit rotation operator $R_y(s_i)$;

$$|M\rangle_u|S\rangle_d = |M\rangle_u\left[\otimes_{i=1}^{N} R_y(s_i)|M\rangle_d\right] = \left[\otimes_{i=1}^{N} R_y(m_i)|\varphi\rangle_u^{(i)}\right]\left[\otimes_{i=1}^{N} R_y(s_i)R_y(m_i)|\varphi\rangle_d^{(i)}\right]. \tag{4}$$

Here, $S = (s_1, s_2, s_3, \ldots, s_N)$ is a rotation angle sequence, $0° \leq s_i \leq 360°$. In addition, S is a private key known only to Alice. Furthermore, we call $|M\rangle_u |S\rangle_d$ to a QMAC state pair.

In the quantum encryption stage, Alice applies quantum encryption $\sigma_{k_{AB}^i} H^{k_H^i}$ to the QMAC state pair $|M\rangle_u |S\rangle_d$ of Equation (4);

$$|M\rangle_u \left[\otimes_{i=1}^N \sigma_{k_{AB}^i} H^{k_H^i} |S\rangle_d \right]. \tag{5}$$

Here, $|M\rangle_u \left[\otimes_{i=1}^N \sigma_{k_{AB}^i} H^{k_H^i} |S\rangle_d \right]$ is an encrypted QMAC state pair, and then she sends it to Bob. This quantum encryption is an essential function for verifying that the entity sending the QMAC pair is Alice and for protecting against forgery.

The rotation angles m_i and s_1 are the elements of the finite discrete variable set. For applying them to real protocols, Alice and Bob must preset the range of the finite discrete variable set and pre-decide how to divide the set range. For example, if Alice and Bob split the rotation angle from $0°$ to $360°$ in intervals of $10°$, then the finite discrete variable set becomes $\{0°, 10°, 20°, \ldots, 350°\}$. Here, the size of the discrete variable set is determined by the performance of the experimental apparatus. Therefore, as the performance of experimental apparatus improves, the size of the discrete variable set increases. Increasing the size of the discrete variable set means that the rotation angle can be subdivided, and this can lead to authenticating more information compared with using the four states of the BB84 protocol. On the other hand, If the performance of the experimental apparatus is poor, the size of the discrete variable set decreases. Then, the rotation angle cannot be subdivided, and information that can be authenticated decreases. Additionally, in this situation, if the communication members use the subdivided rotation angles to such an extent that the experimental apparatus cannot distinguish, detecting the malicious behavior of Eve is impossible.

2.3. Verification Phase

The verification phase is divided into five stages: "quantum decryption", "Bob's encryption", "QMAC recovery", "Bob's decryption", and "swap test". In Stage 1, for quantum decryption, Bob uses secret key sequences K_{AB} and K_H, which were pre-shared with Alice to decrypt the encrypted QMAC state pair $|M\rangle_u \left[\otimes_{i=1}^N \sigma_{k_{AB}^i} H^{k_H^i} |S\rangle_d \right]$ in Equation (5), received from Alice to obtain the QMAC state pair $|M\rangle_u |S\rangle_d$ of Equation (4). In Stage 2, Bob's encryption, Bob generates his own private key sequence $B = (b_1, b_2, \ldots, b_N)$ and re-encrypts quantum state $|S\rangle_d = \otimes_{i=1}^N R_y(s_i)|M\rangle_d$ with it to obtain quantum state $|S'\rangle_d = \otimes_{i=1}^N R_y(b_i)|S\rangle_d$. Then, he sends $|S'\rangle_d$ to Alice while keeping the other quantum message state $|M\rangle_u$. In Stage 3, QMAC recovery, Alice uses her own private key sequence S to apply rotation operator $\otimes_{i=1}^N R_y(-s_i)$ to quantum state $|S'\rangle_d$ and sends quantum state $|S''\rangle_d = \otimes_{i=1}^N R_y(-s_i)|S'\rangle_d$ to Bob. In Stage 4, Bob's decryption, Bob uses his own private key sequence B and applies rotation operator $\otimes_{i=1}^N R_y(-b_i)$ to quantum state $|S''\rangle_d$ to obtain quantum message state $|M'\rangle_d = \otimes_{i=1}^N R_y(-b_i)|S''\rangle_d$. Because the proposed quantum message authentication based on the quantum three-pass protocol operates Alice's private key s_i, there is a need for a method to verify the encrypted QMAC pair described thus far. This is an important element that the proposed protocol can guarantee the origin of quantum message. In addition, to avoid counterfeiting, it is assumed that quantum encryption such as $\sigma_{k_{AB}^i} H^{k_H^i}$ in Equation (5) is applied to Alice and Bob in every process of exchanging quantum states.

In the final stage, Bob performs the swap test [42,45] twice to verify the QMAC state pair. In the first swap test, Bob verifies whether quantum message state $|M\rangle_u$ and quantum message state $|M'\rangle_d$ are the same. If the test result reveals that $|M\rangle_u$ and $|M'\rangle_d$ agree, Bob accepts QMAC state pair $|M\rangle_u |S\rangle_d$ sent by Alice. Otherwise, he does not accept it. In the second swap test, Bob generates quantum state $|M''\rangle$ corresponding to the public bit message sequence M and verifies that it matches quantum message state $|M\rangle_u$ or $|M'\rangle_d$.

If the test result reveals that $(|M''\rangle, |M\rangle_u)$ or $(|M''\rangle, |M'\rangle_d)$ agree, then the integrity of QMAC state pair $|M\rangle_u|S\rangle_d$ is verified completely. For the second swap test, it is noted that the first swap test requires a non-demolition swap test. Figure 2 shows the swap test in the circuit, and the result of inputting

$$|m_i\rangle_u = R_y(m_i)|\varphi\rangle_u^{(i)} \tag{6}$$

and

$$|m_i'\rangle_d = R_y(m_i')|\varphi\rangle_d^{(i)} \tag{7}$$

in the second and third lines of the circuit is expressed as follows:

$$\tfrac{1}{\sqrt{2}}|0\rangle_{ancilla}\left[\tfrac{1}{\sqrt{2}}\left(|m_i\rangle_u|m_i'\rangle_d + |m_i'\rangle_u|m_i\rangle_d\right)\right] + \tfrac{1}{\sqrt{2}}|1\rangle_{ancilla}\left[\tfrac{1}{\sqrt{2}}\left(|m_i\rangle_u|m_i'\rangle_d - |m_i'\rangle_u|m_i\rangle_d\right)\right]. \tag{8}$$

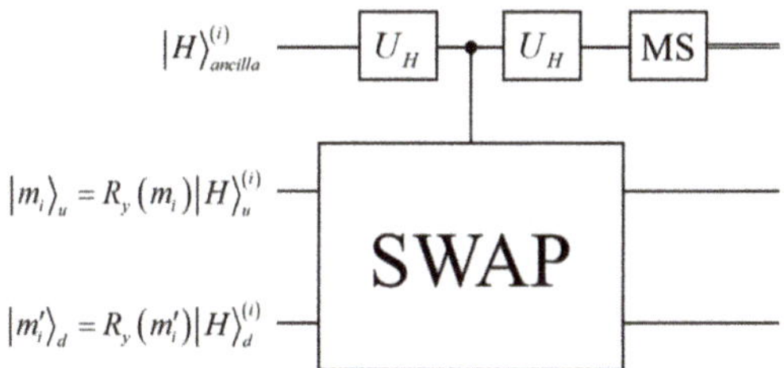

Figure 2. Circuit of the quantum swap test. "SWAP" indicates a swap gate, and U_H represents a Hadamard gate. "MS" represents quantum measurement, and the single lines and the double line represent the quantum channel and classical channel, respectively.

If $|m_i\rangle_u$ and $|m_i'\rangle_d$ agree, the above equation becomes $|0\rangle_{ancilla}\left[\tfrac{1}{\sqrt{2}}\left(|m_i\rangle_u|m_i'\rangle_d + |m_i'\rangle_u|m_i\rangle_d\right)\right]$, which makes the measurement outcome of the ancilla state to always be $|0\rangle$. However, if $|m_i\rangle_u$ and $|m_i'\rangle_d$ do not agree, the measurement outcome becomes $|0\rangle$ with a probability $(1+\varepsilon^2)/2$ or becomes $|1\rangle$ with a probability $(1+\varepsilon^2)/2$, where $\varepsilon = |_d\langle m_i'|m_i\rangle_u|$ and $0 \leq \varepsilon \leq 1$. Therefore, if the swap test result of the measurement is $|1\rangle$, we know that $|m_i\rangle_u$ and $|m_i'\rangle_d$ are different. If the result is $|1\rangle$, we cannot guarantee that $|m_i\rangle_u$ and $|m_i'\rangle_d$ are the same. The parameter ε is determined by the arbitrary quantum state components $|m_i\rangle_u$ of Equation (6) and $|m_i'\rangle_d$ of Equation (7). If the two rotation angles m_i and m_i' are the same, i.e., $m_i = m_i'$, then the value of parameter ε is 1. On the other hand, if the difference between m_i and m_i' is 180°, i.e., $m_i = m_i' \pm 180°$, then the parameter ε is 0. As a result, according to rotation angles m_i and m_i', the parameter ε has a value between 0 and 1, $0 \leq \varepsilon \leq 1$. Further, the probability of failure in the verification phase is the total error probability P_e for N qubits as follows:

$$P_e \leq \otimes_{i=1}^{N}\left[\left(1 + |_d\langle m_i'|m_i\rangle_u|^2\right)/2\right] \tag{9}$$

Therefore, it is expected that the swap test will work well even though the quantum state sequence is finite. Hence, the probability of failure in the verification phase becomes lower, approaching P_e as the size of the quantum state sequence N becomes considerably larger [42,45]. For an arbitrary $|m_i\rangle_u$, a random choice for $|m_i'\rangle_d$ on the $R_y(m_i')$—rotation circle, the average of ε^2 is $1/2$. In this case, the upper bound of the total error probability P_e is $(3/4)^N$. If the size of the quantum state sequence is 15, then the upper bound of the total error probability P_e is only approximately 1.3%. Therefore, it is expected that the swap test will work well even though the quantum state sequence is finite.

3. Security Analysis

3.1. Security of Alice's Private Key

Eve, including Bob, may try to obtain Alice's private key. Especially, as described in Section 2.3, malicious Bob may try to know Alice's private key sequence $S = (s_1, s_2, s_3, \ldots, s_N)$, which consists of the degrees of rotation about $\hat{y}$-axis from $|S\rangle_d = \otimes_{i=1}^{N} R_y(s_i)|M\rangle_d$ in Equation (4). However, the security of Alice's private key sequence S is guaranteed by Holevo's theorem, as follows [19,32]:

$$I(x,S) \leq V(\rho) \leq H(S) \tag{10}$$

Here, $H(S)$ is the Shannon entropy of the sequence of arbitrary rotation angle s_i, $V(\rho)$ is the von Neumann entropy of mixed state ρ that Eve can acquire through the arbitrary measurement of the quantum state $|S\rangle_d = \otimes_{i=1}^{N} R_y(s_i)|M\rangle_d$, and $I(x,S)$ is the mutual information between arbitrary rotation s_i and measurement outcomes x. As we can see in Equation (10), the amount of mutual information about the arbitrary rotation angle sequence S that Bob acquires using measurement outcomes x is limited, and thus, it is impossible for Eve to obtain the information of S. Based on the same principle, the security of Bob's private key sequence $B = (b_1, b_2, b_3, \ldots, b_N)$ is guaranteed.

3.2. Impossibility of Forgery

Many quantum message authentication and signature protocols use quantum encryption implemented by Pauli operators to ensure message integrity and message origin authentication. A QMAC pair (or quantum signature pair), which is composed of a quantum message and an encrypted quantum message, checks the forgery and modulation of the QMAC pair (or quantum signature pair) using a swap test [34]. As described in Section 2.3, Bob validates the original quantum message state $|M\rangle_u$ and the recovered quantum message state $|M'\rangle_d$ from the QMAC state pair of Equation (4) using the swap test. Bob can be sure that $|M\rangle_u$ and $|M'\rangle_d$ are the same quantum state from the outcomes of the swap test. However, it is not known whether they match the original message M. Because of the limitations of this swap test, the proposed protocol can be falsified in two ways.

The first falsification method is that Alice creates a modulated QMAC pair

$$I(x,S) \leq V(\rho) \leq H(S) \tag{11}$$

with the two same quantum states $\left|\widetilde{M}\right\rangle_u$ and $\left|\widetilde{M'}\right\rangle_d$ that do not correspond to the original message M and sends it to Bob. In this case, Bob cannot detect Alice's malicious behaviour even if he verifies that the two quantum states $\left|\widetilde{M}\right\rangle_u$ and $\left|\widetilde{M'}\right\rangle_d$ are the same from the QMAC pair by using the swap test. To prevent this, Alice must disclose message M. Additionally, Bob needs an additional process to validate $|M''\rangle$, which is converted to a quantum state, and $|M\rangle_u$ or $|M'\rangle_d$ by using the swap test.

Second, Eve can try Gao's forgery to apply Pauli operators to a QMAC pair [34,46]. Recently, Gao et al. showed that even if an adversary applies the arbitrary Pauli operator to the QMAC pair (or quantum signature pair), the swap test could not detect it because of the commutation relation of Pauli operators [46]. This is called Gao's forgery, and it can be considered as an existential forgery [34] of modern cryptosystems because it randomly forges QMAC pairs (or quantum signature pairs), which are arbitrary quantum states. The posing of this security problem by Gao et al. was a major turning point in the study of quantum message authentication (or quantum signature) protocols. In 2011, Choi et al. proposed the (I, H)- or (U, V)-type quantum encryption scheme to cope with Gao's forgery [47,48]. In 2013, Zhang et al. pointed out that the encryption scheme of Choi et al. was still insecure against Gao's forgery, and instead they proposed the key-controlled-"I" quantum one-time pad or key-controlled-"T" quantum one-time pad [49,50] as an alternative. The four unitary operators of the controlled-I quantum one-time pad are $W_{00} = (\sigma_x + \sigma_z)/\sqrt{2}$, $W_{01} = (\sigma_y + \sigma_z)/\sqrt{2}$, $W_{10} = (I + i\sigma_x - i\sigma_y + i\sigma_z)/\sqrt{2}$, and $W_{10} = (I + i\sigma_x + i\sigma_y + i\sigma_z)/\sqrt{2}$. However, the encryption scheme of Zhang et al. is not easy to implement with simple hardware. In contrast, we propose a quantum encryption scheme with a single qubit unitary operation by randomly using unitary operator H, which can be easily implemented by controlling wave plates and an authentication protocol. Therefore, the proposed protocol is robust against an existential forgery. Section 4.3 in Ref. [22] shows that unitary operators can be used randomly to prevent Gao's forgery. The detailed implementation of our experimental setup and the testing results of the quantum three-pass protocol and security against Gao's forgery are described in Section 4. Finally, to prevent Gao's forgery in the proposed protocol, the quantum encryption scheme should be applied to all processes in which Alice and Bob exchange quantum states.

3.3. Origin Authentication of Quantum Message

To clarify the origin of the quantum message, the proposed quantum message authentication operates by using not only the secret key pre-shared by Alice and Bob but also Alice's private key. In general, message authentication guarantees the origin of message authentication by using a secret key previously shared by Alice and Bob. At this time, as the user who can create a message authentication code (MAC) pair can be Alice or Bob, the origin of the message may become unclear. On the other hand, in the proposed protocol, Alice generates a QMAC pair $|M\rangle_u |S\rangle_d$ of Equation (4) by using a private key sequence $S = (s_1, s_2, s_3, \ldots, s_N)$ known only to her; thus, the possibility of such a dispute is very low.

4. Experiment Setup and Measurement Results

Figure 3a shows the implementation setup of our proposed quantum message authentication protocol. With this setup, we have experimentally proved that the proposed QMAC is robust against existential forgery. Each stage is implemented with a linear combination of wave plates; that is, the y-axis rotation operator $R_y(\theta)$, the unitary operator H, and the Pauli operators are implemented by combinations of half-wave plates (HWPs) and quarter-wave plates (QWPs). Figure 3b schematically shows a possible forgery attack that Eve can try. Eve can attempt a forgery attack using the same Pauli operators $\sigma_{e_i} = \sigma_{e'_i}$ [46], or she can attempt a forgery attack using different Pauli operators $\sigma_{e_i} \neq \sigma_{e'_i}$ [49,50]. We define these two approaches as an original and improved Gao's Forgeries, respectively. To prevent Gao's forgeries, we need to choose unitary operator H randomly. We explain this in detail at the end of this section.

We assume that Alice and Bob have already pre-shared the secret key sequences in the preparation phase. For the message authentication phase, we implemented message generation, QMAC generation, and quantum encryption using wave plates on Alice's side. To create correlated photon pairs, Type-I spontaneous parametric down-conversion (SPDC) photon pairs were generated in a beta barium borate (BBO) crystal pumped by a multimode diode laser with a 408-nm wavelength. The SPDC photon pairs have the same H-polarization and an 816-nm wavelength. The photon pairs are emitted with a noncollinear angle of 3.3°. One of the photons goes through only the rotation operator for message generation, and the other experiences the sequence of operations from message generation through the quantum encryption scheme with a single qubit unitary operator. Then, they are delivered to Bob. For the verification phase, one photon is kept on Bob's side, and the other photon experiences quantum decryption and Bob's encryption implemented by the wave plate, after which Bob sends it to Alice. Alice then decrypts it by using QMAC recovery. In our experiment, we installed the QMAC recovery stage between Bob's encryption and Bob's decryption for convenience of implementation; it is marked by yellow shading in Figure 3a. Finally, after Bob's decryption, the swap test that verifies the agreement of the two photon sequences is performed using the Hong–Ou–Mandel interferometer. The Hong–Ou–Mandel dip confirms the similarity between the two photons, which is the last step of the implementation of the proposed quantum message authentication protocol.

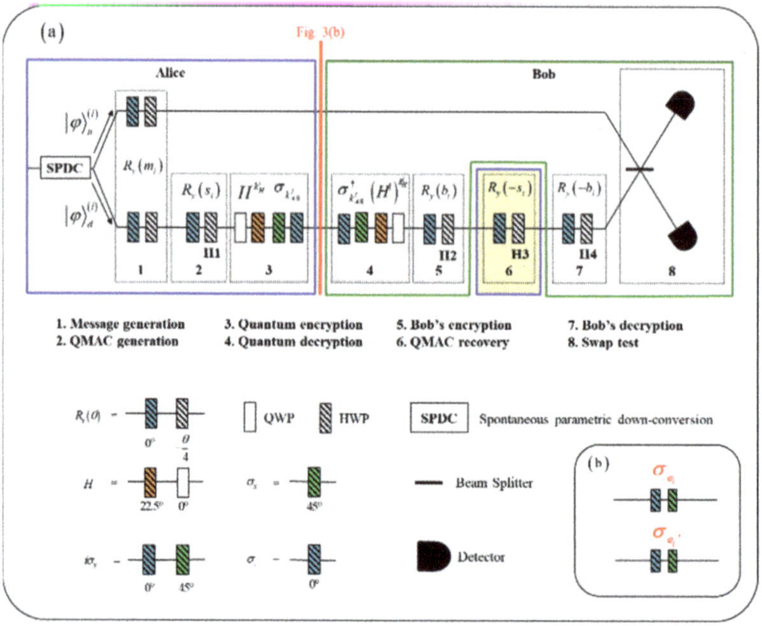

Figure 3. Schematic representation of the experimental setup for the quantum message authentication protocol and an existential forgery. (**a**) Quantum message authentication protocol: the blue box represents Alice's operation, and the green box represents Bob's operation. m_i is the rotation angle that indicates message. k^i_{AB}, k^i_H, s_i, and b_i are the same as in Figure 1. (**b**) Existential forgery: Eve can attempt forgery on the quantum message authentication code (QMAC) state pair using Pauli operators when Alice transmits the encrypted QMAC state pair to Bob.

In other words, the realization of the quantum three-pass protocol, quantum encryption scheme, and the robustness of Gao's forgery can be confirmed by the Hong–Ou–Mandel Dip. Hong–Ou–Mandel interference is the same as the destructive swap test [40]. Because the destructive swap test does not have an ancilla qubit unlike the controlled swap test, the two quantum states that are compared are directly measured and collapsed. For this reason, we performed only the first swap test in the two swap tests shown in Figure 1. To implement the second swap test in Figure 1 using Hong–Ou–Mandel interference, there is a need for more resources (e.g., single photons and wave plates) than the current experimental setup. There are other ways to implement a second swap test by using an experimental controlled swap gate that was recently implemented [51].

We tested the feasibility of our protocol with the experimental setup for the case without Gao's forgery. First, we verified that the quantum three-pass protocol (Figure 3) was working correctly. As shown in Figure 4a, when the half-wave plate H1′s angle $s_i/4$ is $-120°$, the coincidence count reaches its minimum at the half-wave plate H3′s angles $-s_i/4 = 30°$, $120°$ as expected. This indicates that Alice generates the QMAC state by applying rotation operator $R_y(-120°)$ and then uses rotation operator $R_y(-120° \pm \pi n/2)$ to recover the QMAC state, where n is an integer, because the period of the half-wave plate is $\pi/2$. The red plots represent the averages of the coincidence counts over one second. In Figure 4b, we recognize that Bob's encryption and decryption also work well. When the half-wave plate H2′s angle $b_i/4$ is $-60°$, the Hong–Ou–Mandel dip occurs at the half-wave plate H4′s angles $-b_i/4 = 60°$, $150°$. Bob uses rotation operator $R_y(-60°)$ to re-encrypt the QMAC state, and then he decrypts the re-encrypted QMAC state by applying rotation operator $R_y(60° \pm \pi n/2)$, where n is an integer. In Figure 4, the experimental data are the average of 10 measurements per 10 s.

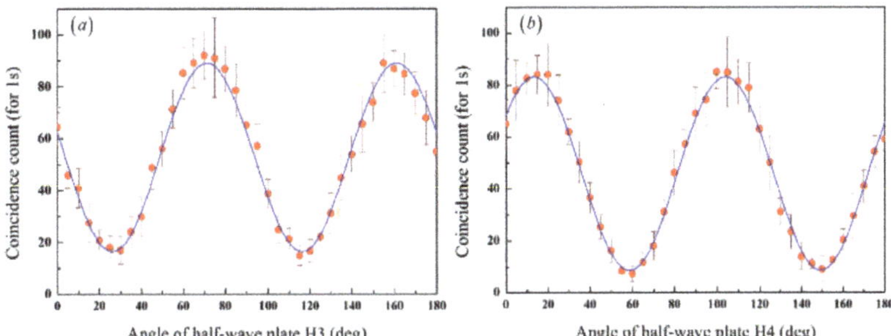

Figure 4. Coincidence counts of the quantum three-pass protocol. The red plots indicate the average of the coincidence counts for one second. The red bars indicate the standard deviation of the coincidence counts for each point. The blue solid line indicates the sine curve fitted to the data. (**a**) Test for QMAC generation and recovery. (**b**) Test for Bob's encryption and decryption.

During this time, the averages of single counts were 27,000 and 27,000, respectively, and coincidence windows are 5 ns; the maximum value of the coincidence counts after accidental coincidences were removed was 127, and the minimum value was 2.

Second, we tested the quantum encryption and decryption. If Alice and Bob are proper users who previously shared secret key sequences K_{AB} and K_H then the quantum message states $|M\rangle_u$ and $|M'\rangle_d$ should be identical. Bob can check the correspondence of these states using the Hong–Ou–Mandel interferometer [38,39]. Figure 4 shows the experimental results for Alice's quantum encryption and Bob's quantum decryption. P_c is the coincidence probability of Hong–Ou–Mandel interference, and $\overline{P}_c = 1 - P_c$ represents the probability of two quantum message states matching. Figure 5a,b represents whether operator H exists or not, respectively. Although theoretically, the red blocks on the diagonal in both cases should be 100%, experimentally they are greater than 82% and 76%, respectively. On the other hand, the blue blocks off the diagonal, when Alice and Bob share different secret keys k^i_{AB} and k^i_H, $|M\rangle_u$ and $|M'\rangle_d$ have different quantum states, and the respective probabilities are less than 41% and less than 46%. Considering that theoretically $\overline{P}_c$ can only have less than 50%, the measurement results prove that our scheme works well. From these results, we can conclude that the encryption operates properly because $\overline{P}_c$ is greater than 76% in the case of the same operators and $\overline{P}_c$ is less than 46% in the case of different operators regardless of the existence of operator H. The above theoretical values are derived from the success probability $\epsilon^2 = \left|_d\langle\psi'_i|\psi_i\rangle_u\right|^2$ of the swap test, with $|\psi_i\rangle_u = U_i R_y(m_i)|0\rangle_u^{(i)}$, $|\psi'_i\rangle_d = U'_i R_y(m_i)|0\rangle_d^{(i)}$, $U_i, U'_i \in \{I, \sigma_x, \sigma_y, \sigma_z, H, \sigma_x H, \sigma_y H, \sigma_z H\}$, and $m_i = 135°$. Errors in the experiment shown in Figure 5 could be due to an inherent error of the swap test, birefringence in the beam splitter, and/or systematic errors in the wave-plate setting [38,39,42,45].

From the measurement results given in Figures 4 and 5, we have demonstrated that our implementation succeeds in realizing the proposed protocol. Although there are some errors due to unavoidable imperfections of the realization, our practical implementation still performs message integrity and message origin authentication successfully only if our protocol is applied to multiple bits sequentially and analyzed statistically.

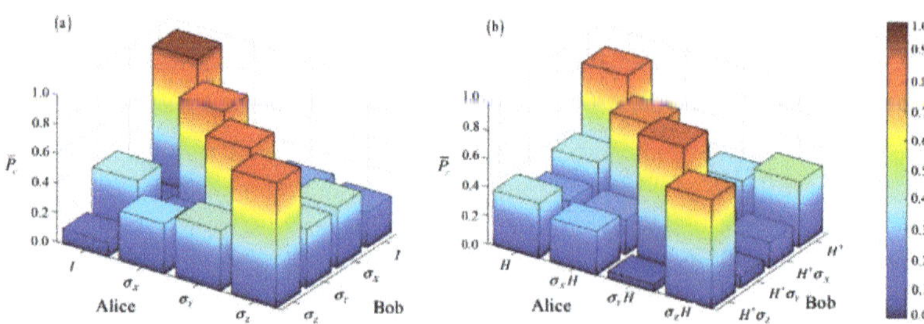

Figure 5. P_c is the coincidence probability of the quantum encryption scheme with a single qubit unitary operator for quantum message authentication. $\overline{P}_c = 1 - P_c$ represents the probability of two quantum message states being matched. In (**a**), $\overline{P}_c$ corresponds to the quantum encryption scheme with a single qubit unitary operator that Alice and Bob can select when secret key k_H^i of Alice and Bob is zero. In (**b**), $\overline{P}_c$ corresponds to every type of quantum encryption with single qubit unitary operator that Alice and Bob can select when secret key k_H^i of Alice and Bob is one. In this experiment, message m_i was set to $135°$.

Gao et al. demonstrated the possibility of existential forgery in the case of quantum message authentication that includes a swap test [34,46,48]. In other words, if the QMAC state pair that Alice generates is not encrypted, Alice cannot detect Eve's intervention. In the quantum encryption $\sigma_{k_{AB}^i} H^{k_H^i}$ in Equation (5), the secret key $k_{AB}^i \in \{00, 01, 10, 11\}$ and $k_H^i \in \{0, 1\}$ correspond to the Pauli operator $\sigma_{k_{AB}^i} \in \{I, \sigma_x, \sigma_y, \sigma_z\}$ and the operator $H^{k_{AB}^i} \in \{I, H\}$ of quantum encryption with a single qubit unitary operator, respectively. The two bits information $e_i \in \{00, 01, 10, 11\}$ corresponds to the Pauli operator $\sigma_{e_i} \in \{I, \sigma_x, \sigma_y, \sigma_z\}$ for Gao's Attack. For example, if $k_{AB}^i = 01$, $k_H^i = 0$, an encrypted QMAC state pair is

$$|M\rangle_u \left[\sigma_{01} H^0 |S\rangle_d\right] = |M\rangle_u \left[\sigma_x |S\rangle_d\right] \tag{12}$$

In addition, the forged QMAC state pair by Eve's Pauli operator $\sigma_{10}(=\sigma_y)$ is

$$\sigma_{10}|M\rangle_u \left[\sigma_{10}\sigma_{01} H^0 |S\rangle_d\right] = \sigma_y |M\rangle_u \left[\sigma_y \sigma_x |S\rangle_d\right] \tag{13}$$

The forged QMAC state pair of Equation (13) transforms into the following state after a decryption process:

$$\sigma_{10}|M\rangle_u \left[\left(H^\dagger\right)^0 \sigma_{01}\sigma_{10}\sigma_{01} H^0 |S\rangle_d\right] = \sigma_y |M\rangle_u \left[\sigma_x \sigma_y \sigma_x |S\rangle_d\right] = \sigma_y |M\rangle_u \left[-\sigma_y |S\rangle_d\right]. \tag{14}$$

Assuming that $|M\rangle_u$ and $|S\rangle_d$ of Equation (14) are the same, Eve succeeded in attacking because the Pauli operator σ_y remained in the first and second qubits of Equation (14). This is the first method to forge the quantum message code or quantum signature pair proposed by Gao et al. [34,46,48].

As another example, if $k_{AB}^i = 01, k_H^i = 1$, an encrypted QMAC state pair is

$$|M\rangle_u \left[\sigma_{01} H^1 |S\rangle_d\right] = |M\rangle_u \left[\sigma_x H |S\rangle_d\right]. \tag{15}$$

The forged QMAC state pair by Eve's Pauli operator $\sigma_{10}(=\sigma_y)$ is

$$\sigma_{10}|M\rangle_u \left[\sigma_{10}\sigma_{01} H^1 |S\rangle_d\right] = \sigma_y |M\rangle_u \left[\sigma_y \sigma_x H |S\rangle_d\right]. \tag{16}$$

The forged QMAC state pair transforms into the following state after a decryption process:

$$\sigma_{10}|M\rangle_u\left[(H^\dagger)^1\sigma_{01}\sigma_{10}\sigma_{01}H^1|S\rangle_d\right] = \sigma_y|M\rangle_u[H^\dagger\sigma_x\sigma_y\sigma_xH|S\rangle_d] = \sigma_y|M\rangle_1[-\sigma_x|S\rangle_2] \quad (17)$$

Despite the assumption that $|M\rangle_u$ and $|S\rangle_d$ of Equation (17) are the same, Eve's attack is unsuccessful. The reason is that the Pauli operators σ_y and σ_x remained in the first and second qubits of Equation (17), respectively. This is the (I, H)-type quantum encryption proposed to overcome Gao's forgery [47]. Zhang et al., however, showed that the (I, H)-type quantum encryption is not secure for improved Gao's forgery [49,50]. We [19,34] overcome the original Gao's forgery [46] or the improved Gao's forgery [49,50] with quantum encryption $\sigma_{k^i_{AB}}H^{k^i_H}$, which randomly uses operator H. Here, the number of all possible cases of quantum encryption $\sigma_{k^i_{AB}}H^{k^i_H} \in \{I,\ \sigma_x,\ \sigma_y,\ \sigma_z,\ H,\ \sigma_xH,\ \sigma_yH,\ \sigma_zH\}$ is 8. Furthermore, except $\sigma_{e_i} = I$, there are three possible ways that Eve can attack with σ_{e_i}. Therefore, there are a total of 24 forgery cases using the Pauli operator σ_{e_i} in the encrypted QMAC state pair $|M\rangle_u\left[\sigma_{k^i_{AB}}H^{k^i_H}|S\rangle_d\right]$ of Equation (5) in the manuscript, Table 1 lists these 24 cases, and Figure 6 shows the results of the experiment with the existential forgery using the Pauli operator for 12 cases in Table 1.

Table 1. A total of 24 forgery cases using the Pauli operator $e_i \in \{\sigma_x,\ \sigma_y,\ \sigma_z\}$ in the encrypted QMAC state pair $|M\rangle_u\left[\sigma_{k^i_{AB}}H^{k^i_H}|S\rangle_d\right]$. Here, $\sigma_{e_i} \in \{\sigma_x,\ \sigma_y,\ \sigma_z\}$, $\sigma_{k^i_{AB}} \in \{I,\ \sigma_x,\ \sigma_y,\ \sigma_z\}$, and $H^{k^i_H} \in \{I,\ H\}$. We assume that the quantum states $|M\rangle_u$ and $|S\rangle_d$ are the same. The yellow shade represents the case where the operator σ_z is not used for quantum encryption or Gao's forgery.

$H^{k^i_H}$	$\sigma_{k^i_{AB}}$	σ_{e_i}	Up(u) Qubit	Down(d) Qubit
$H^0 = I$	$\sigma_{00} = I$	$\sigma_{01} = \sigma_x$	$\sigma_x|M\rangle_u$	$\sigma_x|S\rangle_d$
		$\sigma_{10} = \sigma_y$	$\sigma_y|M\rangle_u$	$\sigma_y|S\rangle_d$
		$\sigma_{11} = \sigma_z$	$\sigma_z|M\rangle_u$	$\sigma_z|S\rangle_d$
	$\sigma_{01} = \sigma_x$	$\sigma_{01} = \sigma_x$	$\sigma_x|M\rangle_u$	$\sigma_x\sigma_x\sigma_x|S\rangle_d = \sigma_x|S\rangle_d$
		$\sigma_{10} = \sigma_y$	$\sigma_y|M\rangle_u$	$\sigma_x\sigma_y\sigma_x|S\rangle_d = -\sigma_y|S\rangle_d$
		$\sigma_{11} = \sigma_z$	$\sigma_z|M\rangle_u$	$\sigma_x\sigma_z\sigma_x|S\rangle_d = -\sigma_z|S\rangle_d$
	$\sigma_{10} = \sigma_y$	$\sigma_{01} = \sigma_x$	$\sigma_x|M\rangle_u$	$\sigma_y\sigma_x\sigma_y|S\rangle_d = -\sigma_x|S\rangle_d$
		$\sigma_{10} = \sigma_y$	$\sigma_y|M\rangle_u$	$\sigma_y\sigma_y\sigma_y|S\rangle_d = \sigma_y|S\rangle_d$
		$\sigma_{11} = \sigma_z$	$\sigma_z|M\rangle_u$	$\sigma_y\sigma_z\sigma_y|S\rangle_d = -\sigma_z|S\rangle_d$
	$\sigma_{11} = \sigma_z$	$\sigma_{01} = \sigma_x$	$\sigma_x|M\rangle_u$	$\sigma_z\sigma_x\sigma_z|S\rangle_d = -\sigma_x|S\rangle_d$
		$\sigma_{10} = \sigma_y$	$\sigma_y|M\rangle_u$	$\sigma_z\sigma_y\sigma_z|S\rangle_d = -\sigma_y|S\rangle_d$
		$\sigma_{11} = \sigma_z$	$\sigma_z|M\rangle_u$	$\sigma_z\sigma_z\sigma_z|S\rangle_d = \sigma_z|S\rangle_d$
$H^1 = H$	$\sigma_{00} = I$	$\sigma_{01} = \sigma_x$	$\sigma_x|M\rangle_u$	$H^\dagger\sigma_xH|S\rangle_d = \sigma_z|S\rangle_d$
		$\sigma_{10} = \sigma_y$	$\sigma_y|M\rangle_u$	$H^\dagger\sigma_yH|S\rangle_d = \sigma_x|S\rangle_d$
		$\sigma_{11} = \sigma_z$	$\sigma_z|M\rangle_u$	$H^\dagger\sigma_zH|S\rangle_d = \sigma_y|S\rangle_d$
	$\sigma_{01} = \sigma_x$	$\sigma_{01} = \sigma_x$	$\sigma_x|M\rangle_u$	$H^\dagger\sigma_x\sigma_x\sigma_xH|S\rangle_d = \sigma_z|S\rangle_d$
		$\sigma_{10} = \sigma_y$	$\sigma_y|M\rangle_u$	$H^\dagger\sigma_x\sigma_y\sigma_xH|S\rangle_d = -\sigma_x|S\rangle_d$
		$\sigma_{11} = \sigma_z$	$\sigma_z|M\rangle_u$	$H^\dagger\sigma_x\sigma_z\sigma_xH|S\rangle_d = -\sigma_y|S\rangle_d$
	$\sigma_{10} = \sigma_y$	$\sigma_{01} = \sigma_x$	$\sigma_x|M\rangle_u$	$H^\dagger\sigma_y\sigma_x\sigma_yH|S\rangle_d = -\sigma_z|S\rangle_d$
		$\sigma_{10} = \sigma_y$	$\sigma_y|M\rangle_u$	$H^\dagger\sigma_y\sigma_y\sigma_yH|S\rangle_d = \sigma_x|S\rangle_d$
		$\sigma_{11} = \sigma_z$	$\sigma_z|M\rangle_u$	$H^\dagger\sigma_y\sigma_z\sigma_yH|S\rangle_d = -\sigma_y|S\rangle_d$
	$\sigma_{11} = \sigma_z$	$\sigma_{01} = \sigma_x$	$\sigma_x|M\rangle_u$	$H^\dagger\sigma_z\sigma_x\sigma_zH|S\rangle_d = -\sigma_z|S\rangle_d$
		$\sigma_{10} = \sigma_y$	$\sigma_y|M\rangle_u$	$H^\dagger\sigma_z\sigma_y\sigma_zH|S\rangle_d = -\sigma_x|S\rangle_d$
		$\sigma_{11} = \sigma_z$	$\sigma_z|M\rangle_u$	$H^\dagger\sigma_z\sigma_z\sigma_zH|S\rangle_d = \sigma_y|S\rangle_d$

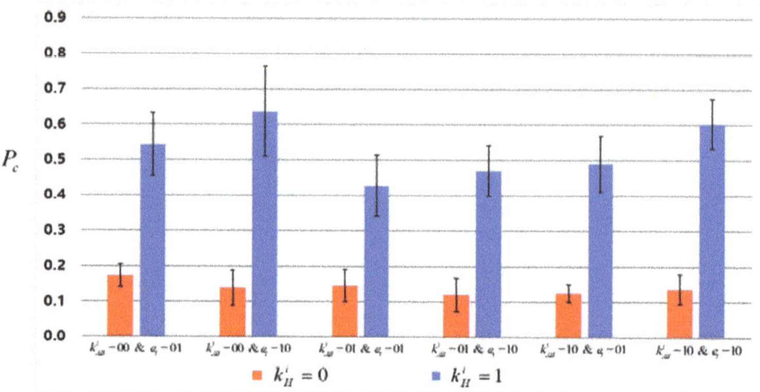

Figure 6. Coincidence probability by existential forgery. Red bars denote the case where Eve attempts original Gao's Forgery when operator H is not used in the quantum encryption scheme $\left(k_H^i = 0\right)$. The blue bars show the case of attempting improved Gao's Forgery when operator H is used in the quantum encryption scheme $\left(k_H^i = 1\right)$. P_c is the coincidence probability. The black bars indicate the standard deviation of the coincidence counts for 1 s. k_{AB}^i is the same as in Figure 1. $e_i \in \{00, 01, 10, 11\}$ corresponds to the Pauli operator $\sigma_{e_i} \in \{I, \sigma_x, \sigma_y, \sigma_z\}$ that Eve uses to attempt Gao's Forgery 1.

5. Conclusions and Discussion

We have proposed a new quantum message authentication protocol including quantum encryption for improving security against an existential forgery. Additionally, a practical implementation of the proposed protocol has been developed and its robustness against existential forgery has been verified experimentally. It consists of wave plates and the Hong–Ou–Mandel interferometer. The measurement results for each function—QMAC generation and recovery, Bob's encryption and decryption, and quantum encryption and decryption—successfully show the feasibility of robustness against Gao's forgeries.

The system loss and the optical channel loss, etc., should be considered when applying our protocol to real implementation. Let us assume that Alice and Bob use the single photon detector with 20% efficiency and are connected by 30-km single-mode fiber with 0.2 dB/km loss. In a result, the total efficiency becomes 0.08% because the qubits are pass through total 100 km, and if the QMAC pairs are generated at 100 MHz, Bob can receive 8×10^4 pairs/s. As we mentioned in Section 2, the size of the quantum state sequence should be more than 15. Therefore, Alice must generate at least 1.9×10^4 QMAC pairs, i.e., $\left(1.9 \times 10^4\right) \times 0.08\% = 15$ that is quite implementable number, and send them to Bob to ensure this accuracy of the swap test.

Our protocol can be used as an arbitrated quantum signature protocol if a trusted center (TC) is added in the communication channel used by Alice and Bob [19]. In addition, if freshness property is added to our protocol, it can be used for quantum entity authentication as well [1,52]. In conclusion, we have proposed the base technology for a complete quantum cryptosystem that provides confidentiality, authentication, integrity, and nonrepudiation.

Author Contributions: M.-S.K. conceived the main idea. M.-S.K. wrote the manuscript. M.-S.K. and Y.-S.K. developed the experimental setup and performed the experiment. M.-S.K., Y.-S.K., J.-W.C., H.-J.Y. and S.-W.H. analyzed the results. S.-W.H. supervised the whole project. All authors reviewed the manuscript. All authors have read and agreed to the published version of the manuscript.

Funding: Korea Institute of Science and Technology (2E30620); National Research Foundation of Korea (2019 M3E4A107866011, 2019M3E4A1079777, 2019R1A2C2 006381); Institute for Information and Communications Technology Promotion (2020-0-00947, 2020-0-00972).

Institutional Review Board Statement: Not applicable.

Informed Consent Statement: Not applicable.

Data Availability Statement: Not applicable.

Conflicts of Interest: The authors declare no conflict of interest.

References

1. Menezes, A.J.; Van Oorschot, P.C.; Vanstone, S.A. *Handbook of Applied Cryptography*; CRC Press: Boca Raton, FL, USA, 1997.
2. Stinson, D.R. *Cryptography*; Chapman & Hall/CRC: Boca Raton, FL, USA, 2006.
3. Bennett, C.H. Quantum crytography. In Proceedings of the International Conference on Computers, Systems & Signal Processing, Bangalore, India, 10–12 December 1984; pp. 175–179.
4. Bennett, C.H. Quantum cryptography using any two nonorthogonal states. *Phys. Rev. Lett.* **1992**, *68*, 3121–3124. [CrossRef]
5. Fuchs, C.A.; Gisin, N.; Griffiths, R.B.; Niu, C.-S.; Peres, A. Optimal eavesdropping in quantum cryptography. I. Information bound and optimal strategy. *Phys. Rev. A* **1997**, *56*, 1163. [CrossRef]
6. Scarani, V.; Acin, A.; Ribordy, G.; Gisin, N. Quantum cryptography protocols robust against photon number splitting attacks for weak laser pulse implementations. *Phys. Rev. Lett.* **2004**, *92*, 057901. [CrossRef]
7. Brassard, G.; Crépeau, C. 25 years of quantum cryptography. *ACM Sigact News* **1996**, *27*, 13–24. [CrossRef]
8. Lo, H.K.; Chau, H.F. Unconditional security of quantum key distribution over arbitrarily long distances. *Science* **1999**, *283*, 2050–2056. [CrossRef] [PubMed]
9. Shor, P.W.; Preskill, J. Simple proof of security of the BB84 quantum key distribution protocol. *Phys. Rev. Lett.* **2000**, *85*, 441–444. [CrossRef]
10. Gisin, N.; Ribordy, G.; Tittel, W.; Zbinden, H. Quantum cryptography. *Rev. Mod. Phys.* **2002**, *74*, 145. [CrossRef]
11. Curty, M.; Santos, D.J. Quantum authentication of classical messages. *Phys. Rev. A* **2001**, *64*. [CrossRef]
12. Barnum, H.; Crépeau, C.; Gottesman, D.; Smith, A.; Tapp, A. Authentication of quantum messages. In Proceedings of the 43rd Annual IEEE Symposium on the Foundations of Computer Science, Vancouver, BC, Canada, 19 November 2002; pp. 449–458.
13. Kang, M.S.; Choi, Y.H.; Kim, Y.S.; Cho, Y.W.; Lee, S.Y.; Han, S.W.; Moon, S. Quantum message authentication scheme based on remote state preparation. *Phys. Scr.* **2018**, *93*. [CrossRef]
14. Zeng, G.; Keitel, C.H. Arbitrated quantum-signature scheme. *Phys. Rev. A* **2002**, *65*, 042312. [CrossRef]
15. Lee, H.; Hong, C.; Kim, H.; Lim, J.; Yang, H.J. Arbitrated quantum signature scheme with message recovery. *Phys. Lett. A* **2004**, *321*, 295–300. [CrossRef]
16. Li, Q.; Chan, W.; Long, D.-Y. Arbitrated quantum signature scheme using Bell states. *Phys. Rev. A* **2009**, *79*, 054307. [CrossRef]
17. Zou, X.; Qiu, D. Security analysis and improvements of arbitrated quantum signature schemes. *Phys. Rev. A* **2010**, *82*, 042325. [CrossRef]
18. Yoon, C.S.; Kang, M.S.; Lim, J.I.; Yang, H.J. Quantum signature scheme based on a quantum search algorithm. *Phys. Scr.* **2014**, *90*, 015103. [CrossRef]
19. Kang, M.-S.; Hong, C.-H.; Heo, J.; Lim, J.-I.; Yang, H.-J. Quantum signature scheme using a single qubit rotation operator. *Int. J. Theor. Phys.* **2015**, *54*, 614–629. [CrossRef]
20. Gottesman, D.; Chuang, I. Quantum digital signatures. *arXiv* **2001**, arXiv:quant-ph/0105032.
21. Clarke, P.J.; Collins, R.J.; Dunjko, V.; Andersson, E.; Jeffers, J.; Buller, G.S. Experimental demonstration of quantum digital signatures using phase-encoded coherent states of light. *Nat. Commun.* **2012**, *3*. [CrossRef] [PubMed]
22. Collins, R.J.; Donaldson, R.J.; Dunjko, V.; Wallden, P.; Clarke, P.J.; Andersson, E.; Jeffers, J.; Buller, G.S. Realization of Quantum Digital Signatures without the Requirement of Quantum Memory. *Phys. Rev. Lett.* **2014**, *113*. [CrossRef] [PubMed]
23. Dunjko, V.; Wallden, P.; Andersson, E. Quantum Digital Signatures without Quantum Memory. *Phys. Rev. Lett.* **2014**, *112*. [CrossRef]
24. Wallden, P.; Dunjko, V.; Kent, A.; Andersson, E. Quantum digital signatures with quantum-key-distribution components. *Phys. Rev. A* **2015**, *91*. [CrossRef]
25. Amiri, R.; Wallden, P.; Kent, A.; Andersson, E. Secure quantum signatures using insecure quantum channels. *Phys. Rev. A* **2016**, *93*. [CrossRef]
26. Donaldson, R.J.; Collins, R.J.; Kleczkowska, K.; Amiri, R.; Wallden, P.; Dunjko, V.; Jeffers, J.; Andersson, E.; Buller, G.S. Experimental demonstration of kilometer-range quantum digital signatures. *Phys. Rev. A* **2016**, *93*. [CrossRef]
27. Yin, H.L.; Fu, Y.; Chen, Z.B. Practical quantum digital signature. *Phys. Rev. A* **2016**, *93*. [CrossRef]
28. Collins, R.J.; Amiri, R.; Fujiwara, M.; Honjo, T.; Shimizu, K.; Tamaki, K.; Takeoka, M.; Sasaki, M.; Andersson, E.; Buller, G.S. Experimental demonstration of quantum digital signatures over 43 dB channel loss using differential phase shift quantum key distribution. *Sci. Rep.* **2017**, *7*. [CrossRef]
29. Yin, H.L.; Fu, Y.; Liu, H.; Tang, Q.J.; Wang, J.; You, L.X.; Zhang, W.J.; Chen, S.J.; Wang, Z.; Zhang, Q.; et al. Experimental quantum digital signature over 102 km. *Phys. Rev. A* **2017**, *95*. [CrossRef]
30. Chan, K.W.C.; El Rifai, M.; Verma, P.; Kak, S.; Chen, Y. Multi-photon quantum key distribution based on double-lock encryption. In Proceedings of the CLEO: QELS_Fundamental Science, San Jose, CA, USA, 10–15 May 2015; p. FF1A.3.
31. Kak, S. A three-stage quantum cryptography protocol. *Found. Phys. Lett.* **2006**, *19*, 293–296. [CrossRef]

32. Nikolopoulos, G.M. Applications of single-qubit rotations in quantum public-key cryptography. *Phys. Rev. A* **2008**, *77*, 032348. [CrossRef]
33. Yang, L.; Wu, L.-A.; Liu, S. Quantum three-pass cryptography protocol. In Proceedings of the Quantum Optics in Computing and Communications, Shanghai, China, 13 September 2002; pp. 106–112.
34. Kang, M.S.; Choi, H.W.; Pramanik, T.; Han, S.W.; Moon, S. Universal quantum encryption for quantum signature using the swap test. *Quantum Inf. Process.* **2018**, *17*. [CrossRef]
35. Massey, J.L.; Omura, J.K. Method and Apparatus for Maintaining the Privacy of Digital Messages Conveyed by Public Transmission. US4567600A, 28 January 1986.
36. Clarke, R.B.M.; Kendon, V.M.; Chefles, A.; Barnett, S.M.; Riis, E.; Sasaki, M. Experimental realization of optimal detection strategies for overcomplete states. *Phys. Rev. A* **2001**, *64*. [CrossRef]
37. Hecht, E.J.I. *Optics*, 4th ed.; Addison-Wesley: San Francisco, CA, USA, 2002; Volume 3, p. 2.
38. Horn, R.T.; Babichev, S.; Marzlin, K.-P.; Lvovsky, A.; Sanders, B.C. Single-qubit optical quantum fingerprinting. *Phys. Rev. Lett.* **2005**, *95*, 150502. [CrossRef]
39. Massar, S. Quantum fingerprinting with a single particle. *Phys. Rev. A* **2005**, *71*, 012310. [CrossRef]
40. Garcia-Escartin, J.C.; Chamorro-Posada, P. Swap test and Hong-Ou-Mandel effect are equivalent. *Phys. Rev. A* **2013**, *87*, 052330. [CrossRef]
41. Curty, M.; Lutkenhaus, N. Comment on "arbitrated quantum-signature scheme". *Phys. Rev. A* **2008**, *77*. [CrossRef]
42. Zeng, G.H. Reply to "Comment on 'Arbitrated quantum-signature scheme'". *Phys. Rev. A* **2008**, *78*. [CrossRef]
43. Riebe, M.; Haffner, H.; Roos, C.F.; Hansel, W.; Benhelm, J.; Lancaster, G.P.T.; Korber, T.W.; Becher, C.; Schmidt-Kaler, F.; James, D.F.V.; et al. Deterministic quantum teleportation with atoms. *Nature* **2004**, *429*, 734–737. [CrossRef] [PubMed]
44. Nielsen, M.A.; Chuang, I.L. *Quantum Computation and Quantum Information*; Cambridge University Press: Cambridge, UK, 2000.
45. Buhrman, H.; Cleve, R.; Watrous, J.; de Wolf, R. Quantum fingerprinting. *Phys. Rev. Lett.* **2001**, *87*. [CrossRef] [PubMed]
46. Gao, F.; Qin, S.J.; Guo, F.Z.; Wen, Q.Y. Cryptanalysis of the arbitrated quantum signature protocols. *Phys. Rev. A* **2011**, *84*. [CrossRef]
47. Choi, J.W.; Chang, K.Y.; Hong, D. Security problem on arbitrated quantum signature schemes. *Phys. Rev. A* **2011**, *84*. [CrossRef]
48. Kang, M.S.; Hong, C.H.; Heo, J.; Lim, J.I.; Yang, H.J. Comment on "Quantum Signature Scheme with Weak Arbitrator". *Int. J. Theor. Phys.* **2014**, *53*, 1862–1866. [CrossRef]
49. Zhang, K.J.; Qin, S.J.; Sun, Y.; Song, T.T.; Su, Q. Reexamination of arbitrated quantum signature: The impossible and the possible. *Quantum Inf. Process.* **2013**, *12*, 3127–3141. [CrossRef]
50. Zhang, K.-J.; Zhang, W.-W.; Li, D. Improving the security of arbitrated quantum signature against the forgery attack. *Quantum Inf. Process.* **2013**, *12*, 2655–2669. [CrossRef]
51. Ono, T.; Okamoto, R.; Tanida, M.; Hofmann, H.F.; Takeuchi, S. Implementation of a quantum controlled-SWAP gate with photonic circuits. *Sci. Rep.* **2017**, *7*. [CrossRef] [PubMed]
52. Hong, C.H.; Heo, J.; Jang, J.G.; Kwon, D. Quantum identity authentication with single photon. *Quantum Inf. Process.* **2017**, *16*. [CrossRef]

Article

Tunability of the Nonlinear Interferometer Method for Anchoring Constructive Interference Patterns on the ITU-T Grid

Kyungdeuk Park, Dongjin Lee and Heedeuk Shin *

Department of physics, Pohang University of Science and Technology (POSTECH), Pohang 37673, Korea;
kyungdeuk@postech.ac.kr (K.P.); dongjin@postech.ac.kr (D.L.)
* Correspondence: heedeukshin@postech.ac.kr

Abstract: Recently, a method of engineering the quantum states with a nonlinear interferometer was proposed to achieve precise state engineering for near-ideal single-mode operation and near-unity efficiency (L. Cui et al., Phys. Rev. A 102, 033718 (2020)), and the high-purity bi-photon states can be created without degrading brightness and collection efficiency. Here, we study the coarse or fine tunability of the nonlinear interference method to match constructive interference patterns into a transmission window of standard 100-GHz DWDM channels. The joint spectral intensity spectrum is measured for various conditions of the nonlinear interference effects. We show that the method has coarse- and fine-tuning ability while maintaining its high spectral purity. We expect that our results expand the usefulness of the nonlinear interference method. The photon-pair generation engineered via this method will be an excellent practical source of the quantum information process.

Keywords: quantum state engineering; nonlinear interferometer; spontaneous four-wave mixing

Citation: Park, K.; Lee, D.; Shin, H. Tunability of the Nonlinear Interferometer Method for Anchoring Constructive Interference Patterns on the ITU-T Grid. *Appl. Sci.* **2021**, *11*, 1429. https://doi.org/10.3390/app11041429

Academic Editor: Maria Bondani

Received: 7 January 2021

Accepted: 1 February 2021

Published: 5 February 2021

Publisher's Note: MDPI stays neutral with regard to jurisdictional claims in published maps and institutional affiliations.

1. Introduction

Modal purity or indistinguishability is an essential factor in achieving high visibility of quantum interference for quantum photonic applications such as quantum teleportation [1] and linear optical quantum computing [2]. High visibility yields high operation fidelity and a high probability of success in quantum information processing using non-classical states of single photons. Among many approaches to obtain non-classical states of photons, spontaneous four-wave mixing (SFWM) has been intensively investigated for the frequency correlated photon-pair generation and heralded single-photon states [3]. However, frequency-correlated photon pairs by the spontaneous parametric process have complicated two-photon states and have a multi-mode nature [4,5]. The multi-mode nature makes photons distinguishable, degrading the quantum interference's visibility. Therefore, spectrally uncorrelated photon pairs with a factorable joint spectral amplitude (JSA) can induce high visibility of quantum interference [4,6] with high indistinguishability and high spectral purity.

The simplest way to obtain spectrally uncorrelated photon-pair is the spectral filtering method with narrow-band filters, but this method can degrade the brightness and photon-number purity due to the optical loss by the filters [7]. The other ways for spectrally uncorrelated photons comes from engineering the dispersion of a parametric medium [8,9]. In spontaneous parametric down-conversion, near unity spectral purity can be achieved with a periodically poled structure [10–12], and SFWM has been tested as well with similar techniques [8,13–17]. While most of the methods are successful to some extent, many sources are expensive to make, not easy to implement, or limited to a specific wavelength range of operation due to strict requirements for dispersion and phase matching [18].

Recently, a new method of engineering the quantum states with nonlinear interference (NLI) is proposed and demonstrated [18–20]. The NLI system consists of two identical nonlinear media and one linear dispersive medium, which is placed in between them. Due to the phase shift induced by the linear dispersive medium, the joint spectral intensity

(JSI) function of bi-photons shows oscillating interference patterns. By adjusting the NLI properties, the authors demonstrated the improved quantum properties of single-photon, spectral correlation, and Hong-Ou-Mandel interference using commercially available optical fibers [19,20]. The author used programmable and tunable filters as the signal and idler filters to reduce noise photons other than the generated photon pair and to have high spectral purity.

Optical fiber systems can guarantee the interoperability between systems and improve their price competitiveness through the standardization of optical communications technology. The standardization of optical frequency channels is necessary for wavelength division multiplexing and is defined by the International Telecommunication Union-Telecommunication standardization sector (ITU-T). Note that the use of customized filters or optical devices raises the entire system's price and makes system development difficult. Using photons on the ITU-T grids makes quantum technologies simple, as most optical devices in current telecommunication systems have frequency channels specified on the ITU-T grids and are commercially available with a reasonable price, high transmittance, narrow bandwidth, well-defined wavelength, etc. The center wavelength of the constructive interference is mainly determined by the length of the linear medium of an NLI system [18–20]. A carefully selected linear medium can make one of the constructive interference modes in a transmission window of DWDM filters without degrading brightness and collection efficiency to obtain high spectral purity. For the case of using single-mode fiber (SMF) as a dispersive medium, only one constructive island can be matched on the ITU grid as Δk_{SMF} of SMF is quadratically dependent on the detuning of pairs from the pump frequency ($\Delta\omega_{s(i)} = |\omega_p - \omega_{s(i)}|$) [18,21]. In order to match all islands on the ITU grids, Δk_{SMF} should be linearly proportional to the detuning, and this requires the replacement of SMF with a device having linear Δk_{SMF} on the detuning [18]. Having pairs over the several ITU-T grids would be helpful for complex multi-photon interference experiments or multi-wavelength channel QKD.

A simple NLI system consists of two dispersion-shifted fiber (DSF) segments as a nonlinear pair generation medium and a section of standard SMF (Corning SMF-28), which is placed between two DSF segments as a linear dispersive medium. The NLI system can be extended by repeatedly adding additional sections of SMF and DSF, and the number of stage (N) is the number of DSF sections in the NLI system. For the case of pumping in the C-band (1530–1560 nm) and using a sufficiently long fiber, C-band photon-pair generation in SMF is negligible for a signal and idler detuning of 400 GHz (~3.2 nm) [21]. Ignoring the propagation loss of fibers and pump chirping, the bi-photon state amplitude through an N-stage NLI system can be calculated by [18],

$$F_{\mathrm{NLI}}(\omega_s, \omega_i) = F_{\mathrm{DSF}}(\omega_s, \omega_i) \times \left[\sum_{n=1}^{N} \exp\{i(n-1)(\Delta k_{\mathrm{DSF}} L_{\mathrm{DSF}} + \Delta k_{\mathrm{SMF}} L_{\mathrm{SMF}})\} \right] \tag{1}$$

where $F_{\mathrm{DSF}}(\omega_s, \omega_i)$ is the JSA at the signal (ω_s) and idler (ω_i) frequency generated by SFWM in a single DSF section, Δk_{DSF} (Δk_{SMF}) is the phase mismatch between signal, idler, and two pump fields in DSF (SMF), L_{DSF} (L_{SMF}) is the length of DSF (SMF), respectively. Each n-th segment of DSF generates an identical bi-photon state to each other, and the phase shift of $\Delta k_{\mathrm{DSF}} L_{\mathrm{DSF}} + \Delta k_{\mathrm{SMF}} L_{\mathrm{SMF}}$ induces the quantum interference in the bi-photon state [18]. From Equation (1), the center wavelength of a constructive interference pattern is tunable by changing Δk_{DSF}, Δk_{SMF}, L_{DSF}, L_{SMF}, and N. The physical parameters such as L_{DSF}, L_{SMF}, and N are variable by changing the length of DSF and SMF and the number of DSF and SMF sections in the NLI system. The optical parameters of Δk_{DSF} and Δk_{SMF} are controllable with the pump wavelength, types of fiber, and temperature.

Here, we investigate the coarse and fine tunability of the nonlinear interference method for creating bi-photon states with high spectral purity. The stimulated-emission-based JSI, which is the absolute square of JSA, measurement technique [22] is used for fast measurements of the JSI spectra, and the measured results are compared to the theoretical prediction, which is calculated using Equation (1). We change the pump wavelength, length of DSF and SMF sections, number of stages, and temperature. The coarse and fine tunability of the NLI method is experimentally demonstrated to control a bright JSI pattern to enter into one of the 100-GHz DWDM filters with the ITU-T channel grid. These results prove that the study in this paper enriches the usefulness and practicality of the NLI method for the efficient photon-pair generation with high spectral purity.

2. Measurement Setup

Figure 1 shows the experimental setup for the stimulated-emission-based JSI measurements. The pump laser is a mode-locked femtosecond (fs) pulse laser (CALMAR FPL-02CTF) with a repetition rate of 18 MHz. Femtosecond pulses are spectrally filtered by a 100-GHz DWDM filter, which has a flat-top spectral shape with about 0.6-nm full-width at half-maximum (FWHM) bandwidth. After filtering, pulses are amplified by an Erbium-Doped Fiber Amplifier (EDFA) and filtered again by DWDM filter to reduce amplified spontaneous emission (ASE) noise from EDFA (not included in Figure 1). The final pump pulse width is about 15 ps and effective bandwidth is about 0.5 nm. The pump peak power (P_p) is about 500 mW, and the seed (or signal) laser is a continuous wave laser with a power of about 10 mW. The lights from the pump and seed lasers are combined by a 200-GHz DWDM filter and injected into an NLI sample. The NLI samples consist of N sections of DSF and N-1 sections of SMF. The length of the DSF and SMF sections is measured using the time-of-flight measurement technique [23]. The polarization states of the pump and seed lights are matched using a polarization controller and an in-line polarizer. The pump light is filtered via a 200-GHz DWDM filter and monitored using an optical power meter to maximize the transmitted power through the in-line polarizer. We discretely vary the wavelength of the seed light while measuring the power of the seed, and generated idler lights using an optical spectrum analyzer. To clearly observe the trend in the shape and position of the JSI spectrum while changing several experimental conditions, we show the normalized spectra. The generated idler power is divided by the input signal power to compensate the system's wavelength-dependent transmittance, and then we normalize it to the maximum value from each measurement's spectrum.

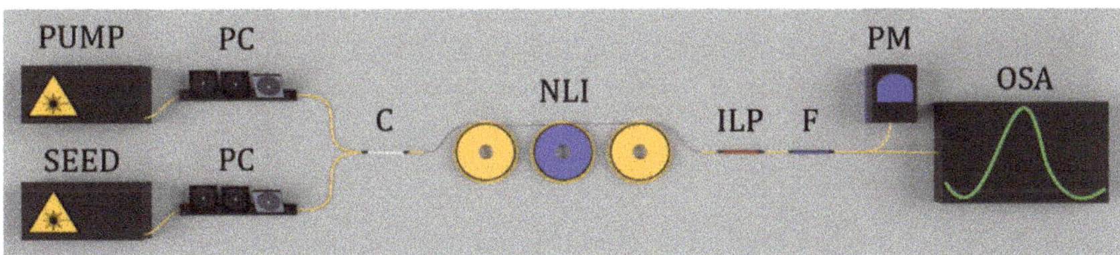

Figure 1. Experimental setup for stimulated-emission-based joint spectral intensity measurement. PUMP: A 15-ps pulse laser with a repetition rate of 18 MHz, PROBE: CW laser for seed and its wavelength is swept during measurement, PC: polarization controller, C: 200 GHz DWDM filter for pump and seed combining, NLI: nonlinear interferometer sample, ILP: in-line polarizer, F: 200 GHz DWDM filter for pump filtering, PM: power meter for pump power monitoring, OSA: optical spectrum analyzer for measuring seed power and generated signal spectrum.

The spectral purity of bi-photons might be changed according to the pump bandwidth (σ_p), but under our experimental conditions, the FWHM pump bandwidth is fixed. Since the pump bandwidth is narrower than that of 100-GHz DWDM filters, we selected the 100-GHz DWDM filters as the target filters of signal/idler photons. In addition, since Raman noise photons increase further away from the pump frequency [24], we decided that 100-GHz DWDM filters, whose center frequency spacing from the pump is ±400 GHz, were the target filters for reducing the Raman noise through the NLI optical fiber system.

3. Results

3.1. Pump Wavelength

First, we test the effect of the pump wavelength (λ_p) on the NLI patterns. When we change the pump wavelength, Δk_{DSF} and Δk_{SMF} are varied, simultaneously. The NLI system under this test consists of two 100-m DSF sections and a 50-m SMF section located between two DSF sections. Figure 2a–c show the measured JSIs for various pump wavelengths, λ_p = [(a) 1550.92 nm (ITU ch.33), (b) 1555.75 nm (ITU ch.27), (c) 1560.61 nm (ITU ch.21)]. The patterns show the constructive and destructive interference patterns as an island arc along the diagonal dashed white line as seen in Figure 2a, and the first bright island ($m = 1$) is about 400 GHz away from the pump frequency. The JSI spectrum in Figure 2a–c looks very similar to each other for various λ_p's, but as seen in Figure 2d, the normalized diagonal JSI spectrum (like the dotted white line in Figure 2a) against the frequency detuning from the pump frequency shows the slow JSI shift towards λ_p as λ_p increases.

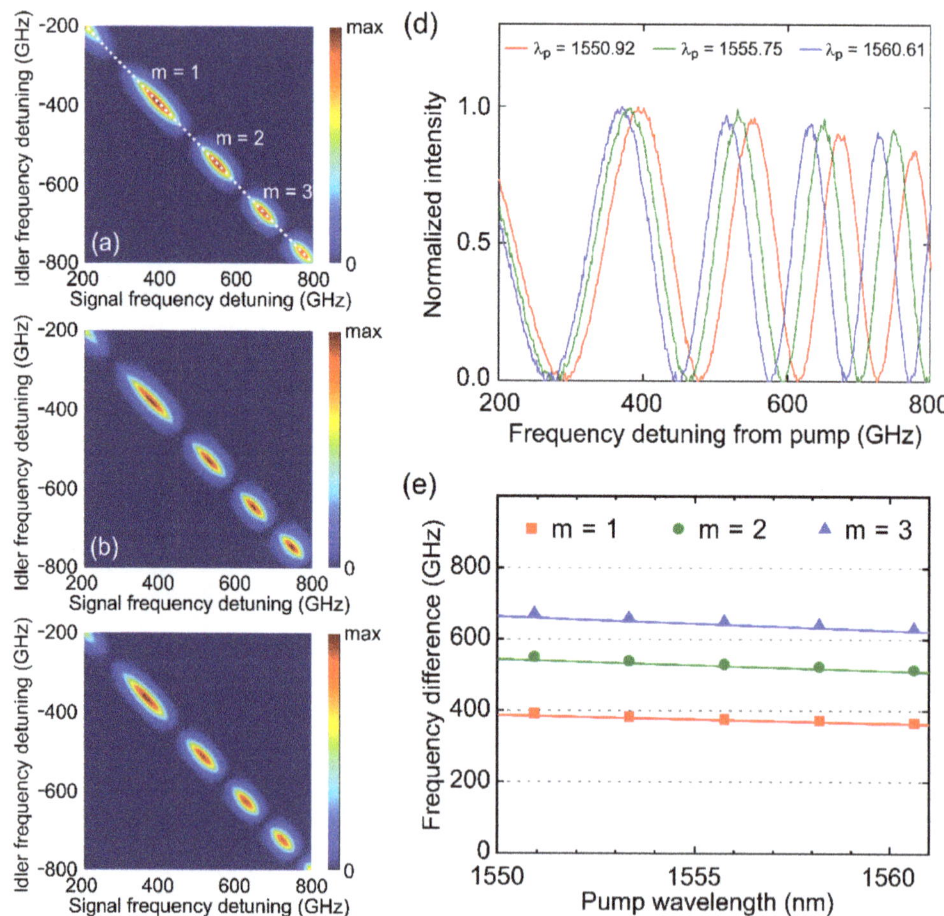

Figure 2. The measured joint spectral intensity (JSI) for various pump wavelengths, (**a**) 1550.92 nm, (**b**) 1555.75 nm, and (**c**) 1560.61 nm. (**d**) The plot of the diagonal line (white dotted line in (**a**)). (**e**) Peak frequency against the pump wavelength for m = 1, 2, and 3. Lines are theoretical predictions extracted from the calculated JSI by using Equation (1).

Figure 2e shows the difference between the pump center frequency and the peak frequency of the first, second, and third islands (m = 1, 2, and 3, respectively) against λ_p. The lines are the theoretical predictions extracted from the calculated JSI using Equation (1). In the JSI calculation, the zero group-velocity-dispersion (GVD) wavelength of DSF ($\lambda_{0,\text{DSF}}$) is 1555.5 nm, and its dispersion slope is 71.5 s/m^3. The zero GVD wavelength of SMF ($\lambda_{0,\text{SMF}}$) is known to be about 1300 nm, and the dispersion of SMF at 1550 nm used in the calculation is 18×10^{-6} s/m^2 with a dispersion slope of 53.3 s/m^3. As seen in Figure 2e, the measured peak shift rates with increasing λ_p are approximately -2.5 GHz/nm, -3.6 GHz/nm, and -4.3 GHz/nm for the interference mode number m = 1, 2, and 3, respectively. As the mode number, m, increases, the peak frequency moves faster towards the pump frequency with increasing λ_p, and the measured results match well with the theoretical predictions. This slow frequency shift yields the fine tunability of the center frequency of islands.

3.2. Length of Nonlinear Pair Generation Medium

Second, we test the effect of the DSF length on the NLI patterns. In references [18–20], the authors assume that the wave vector mismatch goes to zero to ensure the satisfaction of the phase matching conditions. This study, however, considers the case where the wave vector mismatch is small but not negligible. In this case, we expect fine changes in NLI with varying the DSF length.

The NLI system in this test consists of two equal-length DSF sections of various lengths and a 50-m SMF section located between two DSF sections, and λ_p is fixed at 1550.92 nm. Figure 3a–c show the measured JSIs for various DSF lengths (L_{DSF}), and Figure 3d is the normalized diagonal JSI spectra for various DSF lengths against the frequency detuning from the pump frequency. As seen in Figure 3d, the peak frequency gets away from the pump frequency as L_{DSF} becomes longer. Figure 3e is the difference between the pump center frequency and the peak frequency of the first, second, and third islands (m = 1, 2, and 3, respectively) against the DSF length. The frequency shift rates with increasing L_{DSF} are approximately 0.11322 GHz/m, 0.11953 GHz/m, and 0.14266 GHz/m for m = 1, 2, and 3, respectively. The lines are the theoretical predictions using the identical parameters as in Section 3.1 except the DSF lengths and the pump wavelength. The frequency difference slowly grows as increasing L_{DSF} if we consider the phase shift induced in DSF ($\Delta k_{\text{DSF}} L_{\text{DSF}}$) (solid lines in Figure 3e), meanwhile, the difference is constant if we neglect the DSF phase shift (dashed lines). The measured results match well with the theoretical predictions with the phase shift in DSF. This slow frequency shift may induce the fine-tuning ability of the center frequency of islands.

In addition, Figure 3d shows the envelope changes of the pair-generation spectrum for different DSF lengths. It is known that the pair-generation rate is proportional to the square of the SFWM medium length. Recently, we investigate the length dependence of the pair-generation bandwidth, showing that the longer the SFWM length is, the narrower the spectrum bandwidth is [21]. Therefore, a long DSF nonlinear medium can have a large maximum generation rate but a rapidly degraded generation rate at a frequency away from the pump frequency, as seen in Figure 3d. We should carefully select the DSF length for balancing both the bandwidth and pair-generation rate even though the L_{DSF} is the selectable parameter for the fine-tuning of peak frequency.

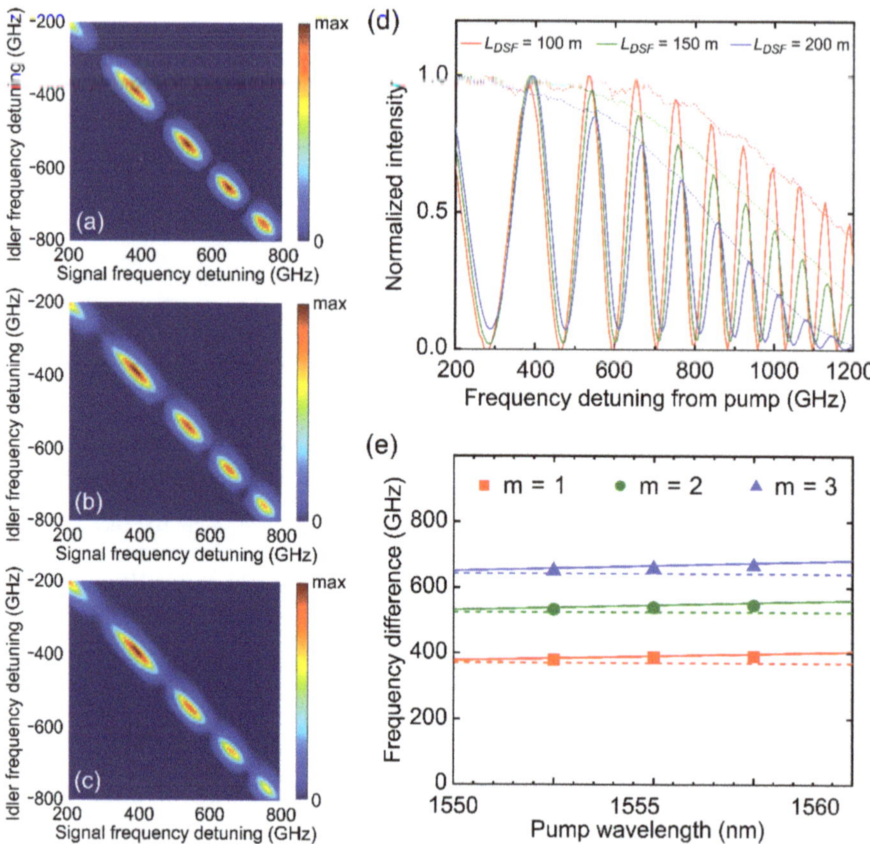

Figure 3. The measured JSI for various dispersion-shifted fiber (DSF) lengths, (**a**) 100 m, (**b**) 150 m, and (**c**) 200 m. (**d**) The plot of the diagonal lines. Dotted lines are measured single stage ($N = 1$, non-interference) JSI case with same length. (**e**) Peak frequency against the DSF length for $m = 1, 2$, and 3. Lines are theoretical predictions extracted from the calculated JSI by using Equation (1) with (solid lines) and without (dashed lines) the phase shift induced in DSF ($\Delta k_{\mathrm{DSF}} L_{\mathrm{DSF}}$).

3.3. Length of Linear Dispersive Medium

The change of L_{SMF} causes dramatic effects on the NLI patterns. In this test, λ_p is fixed (1550.92 nm), and L_{DSF} is 100 m. Figure 4a–d show the measured JSI for various SMF lengths. The spectra right and above the JSI in Figure 4a indicate the transmittance spectra of idler and signal filters, respectively. The filter type used in this study is the DWDM filters whose center frequencies are detuned by 400 GHz from the pump frequency, and its 3-dB bandwidth is about 0.6 nm (~75 GHz). Insets in Figure 4a–d show the JSI islands after passing through the signal/idler filters for $m = 1, 2, 3$, and 4, respectively. Note that the $m = 0$ islands in Figure 4b–d is not shown as the $m = 0$ islands end before 200 GHz. Figure 4e shows the differences between the pump center frequency and the peak frequency for $m = 1, 2, 3$, and 4 islands against the SMF length. The maximum island frequency gets closer as the SMF length becomes longer. The solid curved lines are the theoretical predictions using the identical parameters as in Section 3.1 except the DSF lengths and SMF lengths with the phase shift induced in DSF. The frequency difference decreases rapidly as increasing L_{SMF}. The measured results match well with the theoretical predictions. This fast frequency shift yields the coarse-tuning ability of the center frequency of islands.

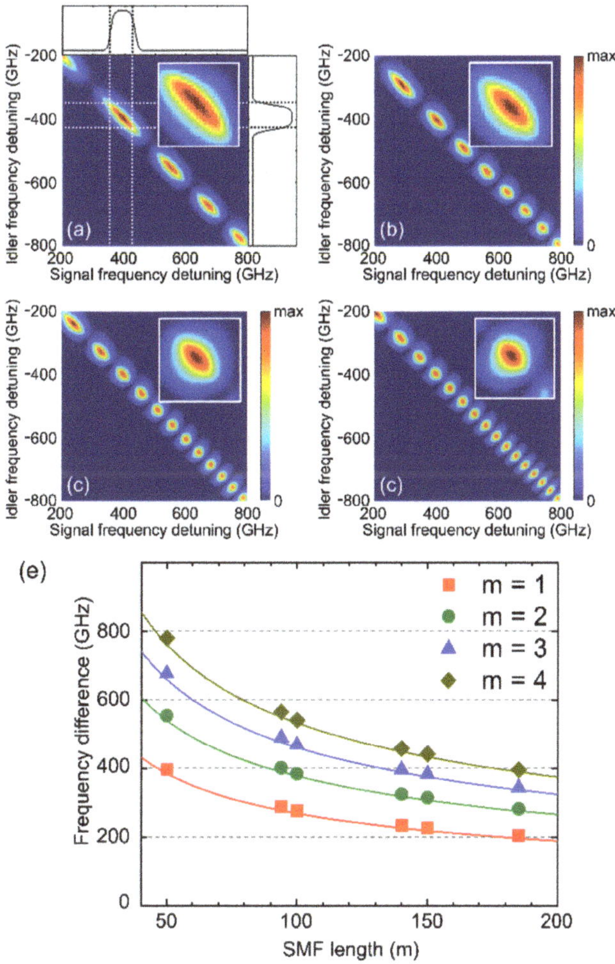

Figure 4. The measured JSI for various SMF lengths, (**a**) 50 m, (**b**) 94 m, (**c**) 140 m, and (**d**) 185 m. Spectra right side and above JSI in (**a**) are signal/idler filters (100-GHz DWDM filter (3dB-bandwidth ~80 GHz)). Inset in each JSI indicates the filtered JSI after the signal/idler filters within filter bandwidth. (**e**) Frequency difference against the SMF length for m = 1, 2, 3, and 4. Lines are theoretical predictions extracted from the calculated JSI by using Equation (1).

The maximum island frequency can be centered at the center of the 100-GHz DWDM filter frequency detuned by 400 GHz from the pump frequency if the condition, $\Delta k_{\mathrm{DSF}}L_{\mathrm{DSF}} + \Delta k_{\mathrm{SMF}}L_{\mathrm{SMF}} = 2\,m\pi$, can be satisfied [18]. The first, second, third, and fourth islands can be within the signal/idler filter transmittance bandwidth if the SMF length is about 50 m, 90 m, 135 m, and 180 m, respectively. The m = 1 island in Figure 4a is elongated diagonally. The JSI bandwidth is wider than the filter bandwidth so that the brightness and collection efficiency, i.e., heralding efficiency, will be degraded. The m = 4 island in Figure 4d, however, is almost round, indicating high spectral purity. The spectral purity (P) can be extracted from the measured data using the Schmidt decomposition method [22]. The extracted p values from the data are p = 0.74, 0.85, 0.95, and 0.91 for inset of (a), (b), (c), and (d), respectively. Therefore, we could obtain the photon pairs with high spectral purity by choosing the proper length of SMF.

3.4. Number of Stages

The change in the number of NLI stages (N) has been demonstrated in reference 18 ($N = 2$) and 19 ($N = 3$) under different NLI conditions, such as the DSF and SMF lengths. Here, we increase the stage number under identical NLI conditions and measure the JSI spectrum and its bandwidth. As seen in Figure 5a–d, varying the stage number does not change the maximum island frequency but the FWHM bandwidth of islands. λ_p is fixed at 1550.92 nm, L_{DSF} is 100 m, and L_{SMF} is 94 m. This condition makes the second island away from the pump frequency by 400 GHz, as seen in Figure 4b. Figure 5a–d are the measured JSI for various Ns. Insets show the JSI filtered by the signal/idler filters. Note that the second island is anchored at the center of the signal and idler DWDM filters even if we change the stage number from $N = 2$ to $N = 5$.

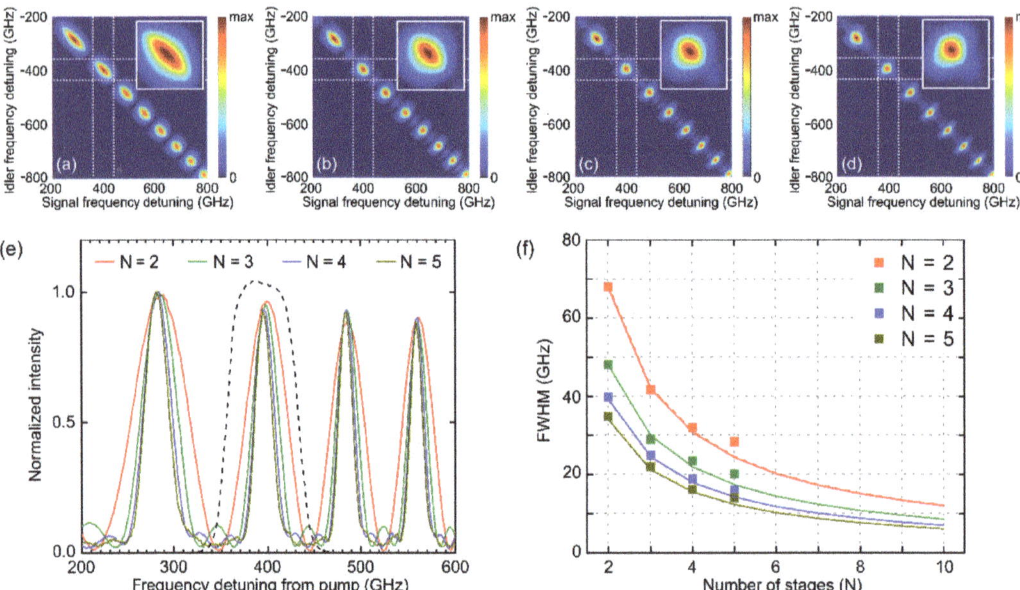

Figure 5. The measured JSI for various number of stage (i.e., number of DSF sample): (**a**) 2, (**b**) 3, (**c**) 4, and (**d**) 5. Inset indicates JSI after signal/idler filter which is 400-GHz detuned 100-GHz DWDM filter (3dB-bandwidth ~ 80 GHz)). (**e**) The plot of the diagonal line. The black dashed line indicates signal/idler filter (flat-top spectral shape). (**f**) Full-Width at Half-Maximum (FWHM) bandwidth of each peak against the number of stages. Lines are theoretical predictions extracted from the calculated JSI by using Equation (1).

Figure 5e is the normalized diagonal JSI spectra for various stage numbers against the frequency detuning from the pump frequency. The black dashed line indicates the signal filter spectrum, and the $m = 2$ island is centered on the filter spectrum. The normalized diagonal JSI spectra in Figure 5e show that the bandwidth of each island gets narrower as the stage number increases, but the peak frequency is stationary. Figure 5f shows the FWHM bandwidth of each island against the stage number. The curved lines are the theoretical predictions with considering the DSF phase shift. The FWHM bandwidth decreases as the stage number increases. The measured data matches well with the theoretical predictions except for the $N = 5$ case. We believe that the non-uniformity of the DSF sections causes this discrepancy, but further study is necessary. The measured p values are 0.82, 0.96, 0.98, and 0.97 for $N = 2, 3, 4,$ and 5, respectively. Therefore, we could obtain the photon pairs with high spectral purity by adding the proper NLI stages.

3.5. Cooling

Finally, we test the temperature effects on the NLI method with the NLI system of $N = 4$ in Section 3.4. Since the silica material of optical fibers can generate noise photons by the Spontaneous Raman Scattering (SpRS) process, the SFWM photon generation medium needs cooling to suppress SpRS. The lower the temperature, the less SpRS occurs [25,26], but cooling the fiber system down to liquid nitrogen temperature is the best cost-effective way. When the optical fiber is cooled down, some properties of optical fibers, such as the zero GVD wavelength, are changed. For the DSF case, $\lambda_{0,\mathrm{DSF}}$ is changed by -4 nm at liquid nitrogen temperature [26], and we assume that $\lambda_{0,\mathrm{SMF}}$ also shifts by -4 nm as the DSF and SMF are made of silica. The dispersion slopes of DSF and SMF are not changed. Figure 6a,b are the theoretical predictions of the JSI at the signal/idler filter frequencies detuned by 400 GHz from the pump frequency, and (c), (d) are the measured JSI. Figure 6a,c are the cases at room temperature with $\lambda_p = 1550.92$ nm. The peak center is placed at around 399.5 GHz in the simulation results and 396 GHz in the experimental results. This peak center shifts if we cool down the temperature to liquid nitrogen temperature. $\lambda_{0,\mathrm{DSF}}$ and $\lambda_{0,\mathrm{SMF}}$ are changed from 1555.5 nm to 1551.5 nm and from 1314 nm to 1310 nm, respectively, without changing the dispersion slope. As seen in Figure 6b,d, the peak center shifts to around 393.5 GHz (-6 GHz changed) in the simulation and 390.5 GHz (-5.5 GHz) in the experiment, respectively. As the shifted island due to cooling is within the signal/idler filter bandwidth and the shape of JSI is almost identical, the spectral purities in the simulation and experiment have almost no change.

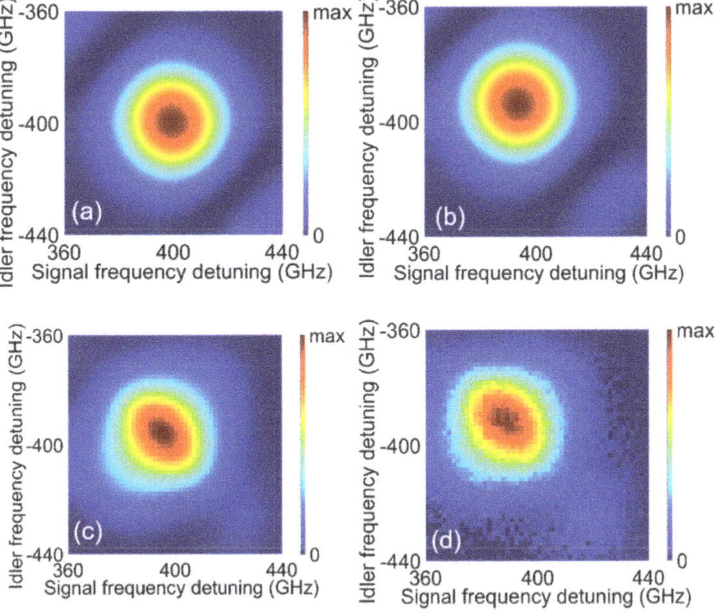

Figure 6. The calculated (**a**,**b**) and measured (**c**,**d**) JSIs at (**a**,**c**) room temperature and (**b**,**d**) liquid nitrogen cooled with $\lambda_p = 1550.92$ nm. The calculation of cooling condition is carried out with the assumption that λ_0 of DSF and SMF are shifted -4 nm and dispersion slopes are not changed. The x- and y-axes are shown ± 40 GHz from filter center which is equivalent to the 3-dB bandwidth of signal/idler filter.

4. Conclusions

Here, we investigate the coarse and fine tunability and cooling effect of the nonlinear interferometer method. The constructive interference islands are controllable by changing the properties of the nonlinear pair generation medium and linear dispersive medium, such as Δk_{DSF}, Δk_{SMF}, L_{DSF}, L_{SMF}, and the number of stages (N). We succeed in adjusting the peak frequency of constructive interference patterns into a transmission window of commercial 100-GHz DWDM filters. As demonstrated in reference 18, 19, and 20, the dominant parameter is L_{SMF}. The peak frequency of a constructive interference island can be coarsely placed near a target wavelength by adjusting the length of SMF. In addition, the pump wavelength and length of DSF can be used for fine-tuning. To match the peak frequency with a DWDM filter center frequency, we should consider the phase shift induced in DSF ($\Delta k_{\mathrm{DSF}} L_{\mathrm{DSF}}$). The selection of L_{DSF} should be carefully considered as not only the peak frequency but also the pair-generation rate and pair-generation spectral bandwidth are also related to L_{DSF}.

Finally, we achieved high spectral purity by selecting an appropriate number of stages (N) and interference mode number (m). The demonstrated methods show the coarse- and fine-tuning ability to match the peak frequency of a constructive interference island and the center frequency of a commercial filter while maintaining high spectral purity. We think that the demonstrated methods in this study expand the usefulness of the NLI method. The generated photon pairs with the engineered quantum state can be an excellent practical source of quantum information processing involving quantum interference.

Author Contributions: Conceptualization, K.P. and H.S.; measurement, K.P.; calculation, D.L.; validation, K.P., D.L. and H.S.; formal analysis, K.P., D.L., and H.S.; writing—original draft preparation, K.P.; writing—review and editing, K.P., D.L. and H.S.; supervision, H.S.; All authors have read and agreed to the published version of the manuscript.

Funding: This research was funded by National Research Foundation of Korea, NRF-2019M3E4A1079780; Korea Institute of Science and Technology's Open Research Program, 2E30620-20-052; Institute for Information & communications Technology Promotion (IITP), No. 2020-0-00947; Affiliated Institute of Electronics and Telecommunications Research Institute (ETRI), 2020-080.

Institutional Review Board Statement: Not applicable.

Informed Consent Statement: Not applicable.

Data Availability Statement: The data presented in this study are available on request from the corresponding author.

Conflicts of Interest: The authors declare no conflict of interest.

References

1. Bouwmeester, D.; Pan, J.; Klaus, M.; Manfred, E.; Harald, W.; Anton, Z. Experimental quantum teleportation. *Nature* **1997**, *403*, 575–579. [CrossRef]
2. Knill, E.; Laflamme, R.; Milburn, G.J. A scheme for efficient quantum computation with linear optics. *Nature* **2001**, *409*, 46–52. [CrossRef]
3. Hong, C.; Mandel, L. Experimental realization of a localized one-photon state. *Phys. Rev. Lett.* **1986**, *56*, 58–60. [CrossRef]
4. Grice, W.P.; Walmsley, I.A. Spectral information and distinguishability in type-II down-conversion with a broadband pump. *Phys. Rev. A* **1997**, *56*, 1627. [CrossRef]
5. Grice, W.P.; U'Ren, A.B.; Walmsley, I.A. Eliminating frequency and space-time correlations in multiphoton states. *Phys. Rev. A* **2001**, *64*, 063815. [CrossRef]
6. Ou, Z.Y.; Rhee, J.K.; Wang, L.J. Photon bunching and multiphoton interference in parametric down-conversion. *Phys. Rev. A* **1999**, *60*, 593. [CrossRef]
7. Brańczyk, A.M.; Ralph, T.C.; Helwig, W.; Silberhorn, C. Optimized generation of heralded Fock states using parametric down-conversion. *New J. Phys.* **2010**, *12*, 063001. [CrossRef]
8. Garay-Palmett, K.; McGuinness, H.J.; Cohen, O.; Lundeen, J.S.; Rangel-Rojo, R.; U'Ren, A.B.; Raymer, M.G.; McKinstrie, C.J.; Radic, S.; Walmsley, I.A. Photon pair-state preparation with tailored spectral properties by spontaneous four-wave mixing in photonic-crystal fiber. *Opt. Express* **2007**, *15*, 14870–14886. [CrossRef] [PubMed]
9. Mosley, P.J.; Lundeen, J.S.; Smith, B.J.; Wasylczyk, P.; U'Ren, A.B.; Silberhorn, C.; Walmsley, I.A. Heralded generation of ultrafast single photons in pure quantum states. *Phys. Rev. Lett.* **2008**, *100*, 133601. [CrossRef]
10. Brańczyk, A.M.; Fedrizzi, A.; Stace, T.M.; Ralph, T.C.; White, A.G. Engineered optical nonlinearity for quantum light sources. *Opt. Express* **2011**, *19*, 55–65. [CrossRef]

11. Dosseva, A.; Cincio, Ł.; Brańczyk, A.M. Shaping the joint spectrum of down-converted photons through optimized custom poling. *Phys. Rev. A* **2016**, *93*, 013801. [CrossRef]
12. Chen, C.; Bo, C.; Niu, M.Y.; Xu, F.; Zhang, Z.; Shapiro, J.H.; Wong, F.N.C. Efficient generation and characterization of spectrally factorable biphotons. *Opt. Express* **2017**, *25*, 7300–7312. [CrossRef]
13. Halder, M.; Fulconis, J.; Cemlyn, C.; Clark, A.; Xiong, C.; Wadsworth, W.J.; Rarity, J.G. Nonclassical 2-photon interference with separate intrinsically narrowband fibre sources. *Opt. Express* **2009**, *17*, 4670–4676. [CrossRef]
14. Cohen, O.; Lundeen, J.S.; Smith, B.J.; Puentes, G.; Mosley, P.J.; Walmsley, I.A. Tailored photon-pair generation in optical fibers. *Phys. Rev. Lett.* **2009**, *102*, 123603. [CrossRef]
15. Smith, B.J.; Mahou, P.; Cohen, O.; Lundeen, J.S.; Walmsley, I.A. Photon pair generation in birefringent optical fibers. *Opt. Express* **2009**, *17*, 23589–23602. [CrossRef]
16. Cui, L.; Li, X.; Zhao, N. Minimizing the frequency correlation of photon pairs in photonic crystal fibers. *New J. Phys.* **2012**, *14*, 123001. [CrossRef]
17. Fang, B.; Cohen, O.; Moreno, J.B.; Lorenz, V.O. State engineering of photon pairs produced through dual-pump spontaneous four-wave mixing. *Opt. Express* **2013**, *21*, 2707–2717. [CrossRef] [PubMed]
18. Cui, L.; Su, J.; Li, J.; Liu, Y.; Li, X.; Ou, Z.Y. Quantum state engineering by nonlinear quantum interference. *Phys. Rev. A* **2020**, *102*, 033718. [CrossRef]
19. Su, J.; Cui, L.; Li, J.; Liu, Y.; Li, X.; Ou, Z.Y. Versatile and precise quantum state engineering by using nonlinear interferometers. *Opt. Express* **2019**, *27*, 20479–20492. [CrossRef] [PubMed]
20. Li, J.; Su, J.; Cui, L.; Xie, T.; Ou, Z.Y.; Li, X. Generation of pure-state single photons with high heralding efficiency by using a three-stage nonlinear interferometer. *Appl. Phy. Lett.* **2020**, *116*, 204002. [CrossRef]
21. Park, K.; Lee, D.; Boyd, R.W.; Shin, H. Telecom C-band Photon-Pair Generation using Standard SMF-28 Fiber. *Opt. Commun.* **2021**, *484*, 126692. [CrossRef]
22. Zielnicki, K.; Garay-Palmett, K.; Cruz-Delgado, D.; Cruz-Ramirez, H.; O'Boyle, M.F.; Fang, B.; Lorenz, V.O.; U'Ren, A.B.; Kwiat, P.G. Joint spectral characterization of photon-pair sources. *J. Mod. Opt.* **2018**, *65*, 1141–1160. [CrossRef]
23. Berkovic, G.; Shafir, E. Optical methods for distance and displacement measurements. *Adv. Opt. Photonics* **2012**, *4*, 441–471. [CrossRef]
24. Park, K.; Lee, D.; Ihn, Y.S.; Kim, Y.-H.; Shin, H. Observation of photon-pair generation in the normal group-velocity-dispersion regime with slight detuning from the pump wavelength. *New J. Phys.* **2018**, *20*, 103004. [CrossRef]
25. Takesue, H.; Inoue, K. 1.5-μm band quantum-correlated photon pair generation in dispersion-shifted fiber: Suppression of noise photons by cooling fiber. *Opt. Express* **2005**, *13*, 7832. [CrossRef] [PubMed]
26. Dyer, S.D.; Stevens, M.J.; Baek, B.; Nam, S.W. High-efficiency, ultra low-noise all-fiber photon-pair source. *Opt. Express* **2008**, *16*, 9966–9977. [CrossRef]

 *applied sciences*

Article

Tailoring Asymmetric Lossy Channels to Test the Robustness of Mesoscopic Quantum States of Light

Alessia Allevi [1,2,*,†] **and Maria Bondani** [2,†]

1   Department of Science and High Technology, University of Insubria, Via Valleggio 11, I-22100 Como, CO, Italy
2   Institute for Photonics and Nanotechnologies, CNR, Via Valleggio 11, I-22100 Como, CO, Italy; maria.bondani@uninsubria.it
*   Correspondence: alessia.allevi@uninsubria.it; Tel.: +39-031-238-6253
†   These authors contributed equally to this work.

Received: 23 November 2020; Accepted: 16 December 2020; Published: 19 December 2020

Abstract: In the past twenty years many experiments have demonstrated that quantum states of light can be used for secure data transfer, despite the presence of many noise sources. In this paper we investigate, both theoretically and experimentally, the role played by a statistically-distributed asymmetric amount of loss in the degradation of nonclassical photon-number correlations between the two parties of multimode twin-beam states in the mesoscopic intensity regime. To be as close as possible to realistic scenarios, we consider two different statistical distributions of such a loss, a Gaussian distribution and a log-normal one. The results achieved in the two cases show to what extent the involved parameters, both those connected to loss and those describing the employed states of light, preserve nonclassicality.

Keywords: mesoscopic quantum states of light; nonclassical photon-number correlations; lossy transmission channels

---

1. Introduction

Since the seminal paper in which Bennett and Brassard dealt with the transmission of quantum states and cryptographic keys through 0.3-m-long free-space air [1], Quantum Communication over long distances has received a lot of interest. In the past two decades, many experiments have been performed using both optical fibers [2–4] and free-space [5–7] channels. Starting from the successful implementation of ground-to-ground atmospheric links [8,9], some most recent experiments have also involved a satellite link [10–12]. Despite all these results, the development of a real global communication network in free-space propagation is still prevented by the atmospheric turbulence, which acts as a temporal and spatial variation of the air refraction index, thus varying the transmittance of the links in a turbulent way. In order to understand the behavior of quantum states of light propagating through atmospheric links, a deep investigation of quantum channels is required [13]. In this respect, quite recent works have introduced different fluctuation loss models capable of accurately describe some experimental results [14,15]. At the same time, it is also important to investigate which kinds of quantum states and nonclassical features are more robust against atmospheric fluctuations and can survive under specific conditions [16]. Until now, most experiments have been implemented at the single-photon level. Recently, we have realized an experiment involving mesoscopic twin-beam states, in which signal and idler were affected by different amounts of loss distributed according to specific statistical distributions [17]. In particular, we have investigated how nonclassicality changes as a function of the mean value of the distributions for fixed values of their standard deviation. Here we face the problem from a different point of view, that is we consider the case in which the standard deviation of the distribution is varied and the mean value of the distribution is fixed. At variance with the previous paper, we deeply analyze the behavior of nonclassicality by choosing worse and worse situations, down to very low and very noisy transmission-efficiency values. For all the chosen values of the parameters, we consider two possible transmittance distributions, namely the log-normal distribution [18,19], which is typically used to describe very turbulent transmission channels, and the Gaussian distribution, which can be exploited to model free-space channels under specific weather conditions [20]. To investigate the nonclassicality of the generated twin beams, which are entangled in the number of photons, we consider the noise reduction factor (R). Indeed, it has been demonstrated that $R < 1$ represents a sufficient criterion for entanglement in the case of bipartite Gaussian states [21]. Actually, many other nonclassicality criteria exist and have been proposed over the years [22–27]. However, not all of them can be easily applied to any experimental situation. Some of them, such as those based on separability criteria, require the full reconstruction of the state under investigation. On the contrary, the noise reduction factor can be calculated from experimental data in a more direct way, also because it can be defined in terms of measurable quantities [28] (see the next section). The obtained results shed light on the different behavior of nonclassicality according to the chosen distribution of the transmission efficiency and on the evolution of such a behavior as a function of the different parameters. To be more exhaustive, in this work we also investigate the role played by the mean number of photons per mode of the employed quantum states in the entanglement degradation process.

Our analysis can give some useful hints not only for the exploitation of quantum states in communication protocols, but also for their application in different contexts, such as for imaging protocols [29–31].

2. Materials and Methods

2.1. Statistical Distributions of the Transmittance Coefficient

The transmission of light through a linear medium can be described by a wavelength- and polarization-dependent transmission coefficient. However, noise effects, such as absorption, depolarization, dephasing and scattering, can occur simultaneously, thus determining a variable transmittance coefficient or, more properly, a statistically-distributed transmittance coefficient. The situation is even more critical when the transmitted light is not represented by a single state but rather by a multipartite one. Indeed, in such a case, the different components of the state can experience different transmission efficiencies. This fact can determine the degradation of the correlations existing among the parties. It is thus crucial to quantify the amount of degradation in order to verify if the employed multipartite state is still useful for applications or not [32]. For instance, in the case of correlated bipartite systems, both classical and quantum, it is possible to get a fair estimation of the occurrence of degradation in terms of the noise reduction factor R, which is defined as

$$R = \frac{\sigma^2(n_1 - n_2)}{(\langle n_1 \rangle + \langle n_2 \rangle)}, \tag{1}$$

in which $\sigma^2(n_1 - n_2)$ is the variance of the distribution of the photon-number difference between the two parties, while $(\langle n_1 \rangle + \langle n_2 \rangle)$ is the shot-noise-level, that is the value of $\sigma^2(n_1 - n_2)$ in the case of coherent states having mean values $\langle n_1 \rangle$ and $\langle n_2 \rangle$.

Typically, the noise reduction factor is used to test the nonclassicality of quantum states of light, $R < 1$ being a sufficient condition for entanglement be expressed in terms of measurable quantities, such as the "detected" number of photons. As an example, in the case of multimode twin-beam states, the measured noise reduction factor reads [33]

$$R = 1 - \frac{2\sqrt{\eta_1 \eta_2}\sqrt{\langle m_1 \rangle \langle m_2 \rangle}}{\langle m_1 \rangle + \langle m_2 \rangle} + \frac{(\langle m_1 \rangle - \langle m_2 \rangle)^2}{\mu(\langle m_1 \rangle + \langle m_2 \rangle)} \tag{2}$$

and the condition $R < 1$ proves the existence of sub-shot-noise correlations between the two parties. In Equation (2), $\langle m_1 \rangle$ and $\langle m_2 \rangle$ are the mean number of detected photons in the two arms of twin beam, μ is the number of modes, and η_1 and η_2 are the detection efficiencies. In Ref. [17] we considered the presence of an asymmetric loss in the two arms of twin beam, and defined $\langle m_1 \rangle = \langle m \rangle = \eta \langle n \rangle$ and $\langle m_2 \rangle = \langle m \rangle t = \eta \langle n \rangle t$ ($t \in (0,1)$). The expression for R modifies as:

$$R = 1 - \frac{2\eta t}{1+t} + \frac{(1-t)^2}{(1+t)} m_\mu,\tag{3}$$

in which $m_\mu = \langle m \rangle / \mu$ is the mean number of photons per mode. As stated above, in realistic situations, the asymmetric transmittance coefficient t is not constant but rather statistically distributed. In this case, by following the procedure presented in Ref. [17], it is still possible to find an analytic expression for the noise reduction factor R, that is

$$R = 1 - \frac{2\eta \langle t \rangle}{1 + \langle t \rangle} + \frac{1 - \langle t \rangle)^2}{1 + \langle t \rangle} m_\mu + (\langle m \rangle + m_\mu) \frac{\sigma^2(t)}{1 + \langle t \rangle}.\tag{4}$$

We notice that in Equation (4) only the first two moments of the statistics of t are involved, namely $\langle t \rangle$ and $\sigma^2(t)$. This makes the investigation of the degradation of entanglement for twin-beam states particularly straightforward once the two moments of the distribution of t are known (or can be calculated). As an example, in Figure 1 we show a 3D plot of R as a function of $\langle t \rangle$ and $\sigma(t)$ for $\eta = 0.145$, $\langle m \rangle = 2.23$ and $m_\mu = 0.04$, which represent typical experimental values.

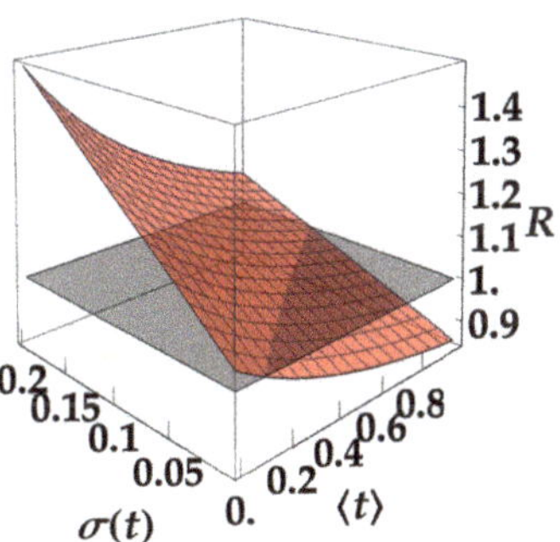

Figure 1. Theoretical expectation, according to Equation (4), of R (red surface) as a function of $\langle t \rangle$ and of $\sigma(t)$ for $\eta = 0.145$, $\langle m \rangle = 2.23$ and $m_\mu = 0.04$. The gray surface at $R = 1$ represents the boundary plane between classical and nonclassical correlations.

As expected, the observation of nonclassical correlations becomes very difficult both at low values of $\langle t \rangle$ and at values of $\sigma(t)$ exceeding a threshold that depends on $\langle t \rangle$.

In the following, we consider two loss distributions that are involved in the propagation process of light through media. In particular, we deal with Gaussian and log-normal distributions. The Gaussian distribution of t can be expressed as

$$P_g(t) = \frac{1}{\sqrt{2\pi}\sigma_0} \exp\left[\frac{-(t - t_0)^2}{2\sigma_0^2}\right],\tag{5}$$

where t_0 and σ_0 are the mean value and the standard deviation for $t \in (-\infty, +\infty)$, while the log-normal distribution is usually defined as

$$P_{\ln}(t) = \frac{1}{\sqrt{2\pi}\sigma_\zeta t} \exp\left[\frac{-[-\zeta + \ln(t)]^2}{2\sigma_\zeta^2}\right],\tag{6}$$

in which $\xi \in \Re$ is the so-called location parameter and $\sigma_\xi > 0$ is the scale parameter. Both parameters are linked to the mean value and the standard deviation of the distribution for $t \in (0, +\infty)$: $t_0 = \exp[\xi + \sigma_\xi^2/2]$ and $\sigma_0^2 = (\exp[2\xi + \sigma_\xi^2])(-1 + \exp[\sigma_\xi^2])$.

For the specific application we are considering, the transmittance coefficient t is limited to the interval (0,1), which means that the distributions in Equations (5) and (6) must be properly normalized in the interval (0,1). The resulting probabilities can be expressed through closed formulas and the same holds for the first two moments of the distributions, $\langle t \rangle$ and $\sigma(t)$.

In order to appreciate the differences between the two distributions, in Figure 2 we show some examples. We observe that, in general, at increasing the standard deviation, the discrepancies between Gaussian and log-normal distributions become more evident, since the log-normal distribution gets more asymmetric, while the Gaussian one remains symmetric. Moreover, comparing the two distributions for a given choice of standard deviation and mean value shows that for small mean values of t_0 (see panels (a) and (b)) and large values of σ_0 (see magenta curves) the log-normal distribution is confined in the region corresponding to small values of t (dashed line), while the Gaussian one is uniformly distributed all over the range (0,1) (solid line). On the contrary, for large mean values of t_0 (see panels (c) and (d)) and large values of σ_0 (see magenta curves) the Gaussian distribution has a longer tail towards small values of t than the log-normal one. At variance with these conditions, in all panels for small values of σ_0 (see black curves) the two distributions are more confined and more similar.

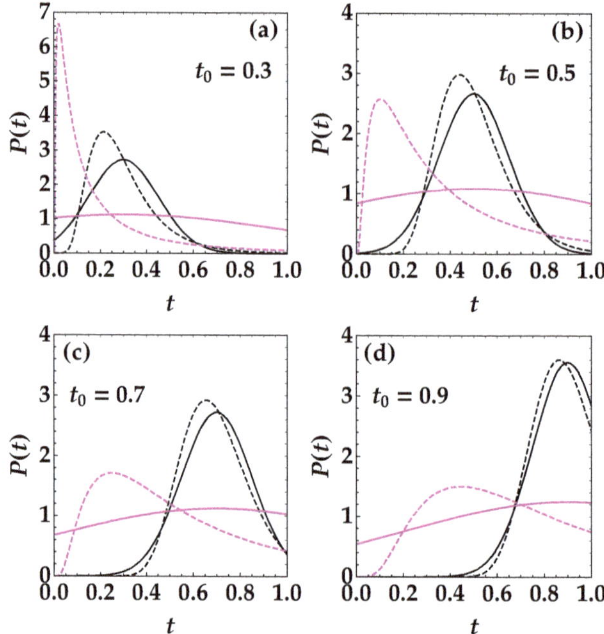

Figure 2. Gaussian (solid lines) and log-normal (dashed lines) probability distributions of t in the interval (0, 1). The four panels (**a–d**) correspond to different choices of the mean value t_0 of the distributions, as indicated in the label on top of each panel. Inside each panel, the different colors correspond to different choices of the standard deviation: $\sigma_0 = 0.15$ is represented in black, while $\sigma_0 = 0.7$ in magenta.

Before presenting the results of our investigation, we summarize the main features of the experimental implementation and explain the method used to prepare the data according to the specific distributions.

2.2. Experimental Setup and Data Preparation

A sketch of the employed experimental setup in shown in Figure 3. The fundamental (at 1047 nm) and the third harmonic (at 349 nm) of a Nd:YLF laser (IC-500, High Q Laser, Rankweil, Austria and 2003) regeneratively amplified at 500 Hz are sent to a β−barium-borate crystal (BBO1, cut angle = 37°, 8-mm long) in order to produce the fourth-harmonic (at 262 nm) field by noncollinear sum-frequency generation. The generated beam is used to pump spontaneous parametric downconversion in a second BBO (BBO2, cut angle = 46.7°, 6-mm long) crystal.

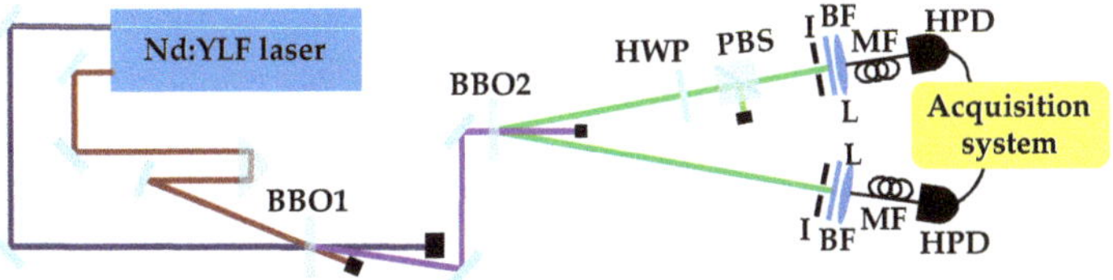

Figure 3. (Color online) Sketch of the experimental setup. See the text for details.

Two twin portions at frequency degeneracy are selected both spatially and spectrally by means of two irises (I) and two bandpass filters (BF), respectively. The filtered portions are then focused into two multimode fibers (600-μm-core diameter) and delivered to two hybrid photodetectors (HPD, mod. R10467U-40, Hamamatsu Photonics, Hamamatsu City, Japan and 2010). These are photon-number-resolving detectors endowed with a partial photon-number-resolving capability and a good linearity up to 100 photons. Each detector output is amplified, synchronously integrated and digitized. As extensively explained in previous papers, by applying a self-consistent method to each detector output it is possible to have access to detected photons and thus to the statistical properties of the measured states [33,34].

In order to introduce a variable transmittance coefficient (from 0 to 1) only in one arm of the twin-beam state, a half-wave plate (HWP) followed by a polarizing cube beam splitter (PBS) is inserted in that arm. During the measurements, the half-wave plate is rotated in steps of 2° and 50,000 acquisitions are saved for each angle value.

To build a given distribution $P(t)$ of the transmission coefficient, for each measured value t_i in the interval (0,1) we select a dataset of $P(t_i)$ elements on both arms and join all the chosen datasets preserving the correspondence of the single data in the two arms. In such a way, the statistics of light in the arm without variable transmittance remains unchanged, while that on the other arm gets modified. This determines a degradation of the nonclassical photon-number correlations between the two arms, which can be quantified by evaluating the noise reduction factor.

Note that this procedure allows us to check, starting from the same datasets, the effect of different distributions of t by simply changing $P(t)$ and choosing different values of t_0 and σ_0.

3. Results

At variance with Ref. [17], in which we focused our attention on the possibility of keeping observing nonclassicality in the presence of an asymmetric loss between the two parties of the twin-beam states, here we aim at finding the limits imposed by the parameters that describe the transmittance statistics, both for the Gaussian and the log-normal distributions. According to Equation (4), the noise reduction factor is a function of the first two moments of the distribution of t, independent of the considered distribution. This means that by choosing the same mean value $\langle t \rangle$ and the same standard deviation $\sigma(t)$, the Gaussian and the log-normal distributions act in the same way. On the contrary, if we consider the values of t_0 and σ_0 over the entire domains, we expect different results. To better emphasize this point, in panel (a) of Figure 4 we show, as contour plot, the difference $\langle t \rangle_{\mathrm{GAUSS}} - \langle t \rangle_{\mathrm{LOG}}$ between the mean value $\langle t \rangle$ of the normalized Gaussian distribution and that of the normalized log-normal distribution as a function of t_0 and σ_0, while in panel (b) of Figure 4 we do the same for the standard deviation $\sigma(t)$, namely we plot the difference $\sigma(t)_{\text{GAUSS}} - \sigma(t)_{\text{LOG}}$. In panel (a) (panel (b)), the pink region corresponds to values of $\langle t \rangle_{\text{GAUSS}} - \langle t \rangle_{\text{LOG}}$ ($\sigma(t)_{\text{GAUSS}} - \sigma(t)_{\text{LOG}}$) larger than 0, the gray region to values lower than 0, and the black-dashed line dividing the pink region from the gray one corresponds to the condition $\langle t \rangle_{\text{GAUSS}} - \langle t \rangle_{\text{LOG}} = 0$ (panel (a)), and to the condition $\sigma(t)_{\text{GAUSS}} - \sigma(t)_{\text{LOG}} = 0$ (panel (b)). As it can be noticed by comparing the two panels, it is not possible to find a set of t_0 and σ_0 values corresponding to equal values of $\langle t \rangle$ and $\sigma(t)$ for the two distributions, unless one considers sparse choices of t_0 and σ_0 for $\sigma_0 < 0.05$.

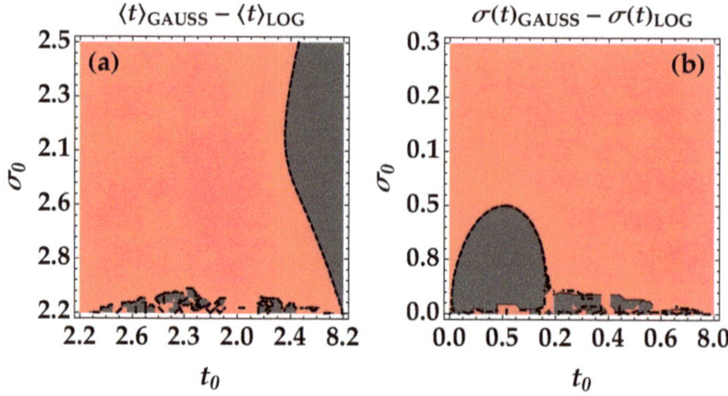

Figure 4. Panel (a): Contour plot of the difference $\langle t \rangle_{\text{GAUSS}} - \langle t \rangle_{\text{LOG}}$ of the Gaussian and log-normal distributions defined over the interval (0,1) as a function of the mean value t_0 and of the standard deviation σ_0 of the same distributions defined over the entire domain. Panel (b): The same as (a) for the difference $\sigma(t)_{\text{GAUSS}} - \sigma(t)_{\text{LOG}}$. In both panels, the black-dashed line corresponds to the condition in which the difference is equal to 0, the pink region to values larger than 0 and the gray region to values lower than 0.

For the above reasons, a direct comparison between the distributions could be not essential. On the contrary, it can be more interesting, for each one of the two distributions separately, to investigate the survival of nonclassical correlations as a function of its first two moments.

In particular, it is straightforward to deal with the expressions defined over the entire domain and their corresponding moments by considering the link among $\langle t \rangle$ and $\sigma(t)$ in Equation (4) to t_0 and σ_0.

First of all, we consider the case of the Gaussian distribution. In Figure 5 we show the noise reduction factor as a function of the standard deviation σ_0 for different values of t_0. Even if, in general, σ_0 can assume values larger than 1, in our analysis we explored only values in the interval (0,1) since they are sufficient to observe the degradation of nonclassicality. The data, shown as colored dots + error bars, are superimposed to the theoretical fitting curves that can be obtained according to Equation (4), in which we fix the values of $\langle m \rangle$ and η and leave m_μ and $t_{0,\text{FIT}}$ as fitting parameters. In particular, we set $\langle m \rangle = 2.23$, which is the maximum value of the mean number of photons detected in the variable arm, and $\eta = 0.145$, which is the quantum efficiency of the detection chain obtained as $\eta = 1 - R_{MIN}$, being R_{MIN} the value of the noise reduction factor corresponding to $t = 1$. In the fitting procedure, we leave $t_{0,\text{FIT}}$ as a free fitting parameter to take into account the possibility that the preparation of the distribution of t is not ideal due to the discrete values of t at our disposal. By inspecting panel (a) of Figure 5, we notice that the smaller the mean value the more difficult the observation of nonclassicality. Indeed, for $t_0 = 0.4$ only small values of σ_0 make the condition $R < 1$ possible. On the contrary, for $t_0 = 0.9$ the chance to observe nonclassicality is higher.

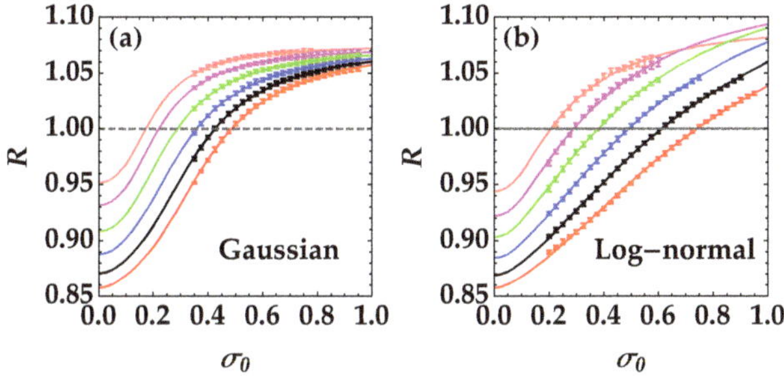

Figure 5. Noise reduction factor as a function of the standard deviation σ_0 of the Gaussian (panel (**a**)) and log-normal (panel (**b**)) distributions defined over the entire domain for different values of t_0. The experimental data are shown as colored dots + error bars: pink dots refer to $t_0 = 0.4$, magenta dots to $t_0 = 0.5$, green dots to $t_0 = 0.6$, blue dots to $t_0 = 0.7$, black dots to $t_0 = 0.8$, and red dots to $t_0 = 0.9$. The corresponding theoretical fitting functions are plotted as colored curves with the same color choice. In the fitting procedure we fixed $\langle m \rangle = 2.23$ and $\eta = 0.145$ and we left $t_{0,\mathrm{FIT}}$ and m_μ as fitting parameters. The values of such parameters are reported in Tables 1 and 2 together with the χ^2 per degree of freedom.

The behavior exhibited by the noise reduction factor in the case of the log-normal distribution, shown in panel (b) of Figure 5, is quite similar. Indeed, also in this case, observing the nonclassicality is particularly difficult for small values of t_0. However, it is interesting to notice that, at variance with the Gaussian distribution, for large values of t_0 the values of R linearly increase as a function of σ_0, while for small values the behavior is more similar to a sigmoid. The different behavior of R for different choices of t_0 is due to the nontrivial and asymmetric shape of the log-normal distribution at different mean values, as already shown in Figure 2. The fitting parameters corresponding to the data shown in Figure 5 are summarized in Table 1 for the Gaussian distributions and in Table 2 for the log-normal ones. In particular, we note that the fitted values of m_μ are almost constant in the case of Gaussian distributions, while they change in the case of log-normal ones.

Table 1. Values of the fitting parameters $t_{0,\mathrm{FIT}}$ and m_μ of the noise reduction factor as a function of σ_0 for $\langle m \rangle = 2.23$ and $\eta = 0.145$ in the case of Gaussian distributions of t. In the last column the χ^2 per degree of freedom is shown.

t_0	$t_{0,\mathrm{FIT}}$	m_μ	χ^2_ν
0.4	0.460 ± 0.002	0.2171 ± 0.0007	0.018
0.5	0.523 ± 0.002	0.2115 ± 0.0007	0.060
0.6	0.616 ± 0.002	0.2115 ± 0.0008	0.054
0.7	0.721 ± 0.002	0.213 ± 0.001	0.118
0.8	0.839 ± 0.003	0.217 ± 0.002	0.198
0.9	0.964 ± 0.003	0.226 ± 0.002	0.463

Table 2. Values of the fitting parameters $t_{0,\mathrm{FIT}}$ and m_μ of the noise reduction factor as a function of σ_0 for $\langle m \rangle = 2.23$ and $\eta = 0.145$ in the case of log-normal distributions of t. In the last column the χ^2 per degree of freedom is shown.

t_0	$t_{0,\mathrm{FIT}}$	m_μ	χ^2_ν
0.4	0.369 ± 0.007	0.077 ± 0.004	0.418
0.5	0.508 ± 0.007	0.126 ± 0.006	0.784
0.6	0.618 ± 0.004	0.161 ± 0.005	0.423
0.7	0.734 ± 0.002	0.189 ± 0.003	0.309
0.8	0.852 ± 0.002	0.213 ± 0.003	0.346
0.9	0.964 ± 0.003	0.226 ± 0.006	0.697

4. Discussion

The different behavior of the mean number of photons per mode for the two distributions can be ascribed to the fact that, when the mean number of photons is varied in one arm of the twin beam, also the number of modes changes. Indeed, the number of modes we obtain from the first two moments of the light statistics is an "effective" one describing the multimode state as the tensor product of μ equally-populated single-mode states. This is just a useful approximation since in the real experiment the different modes are differently populated [35]. For this reason, the attenuation of light by a filter (the HWP followed by the PBS is equivalent to such a condition) can vary the number of effective modes since some of the real modes can be so attenuated to go below the detection threshold. In particular, we experimentally observed that the number of effective modes monotonically increases at increasing the values of the mean number of photons. The variability of the number of modes is amplified when the data are combined according to specific distributions of t. As discussed above, this is more visible in the case of log-normal distribution due to its nontrivial shape at different mean values.

To better investigate the role played by the mean number of photons per mode in the calculation of the noise reduction factor for the two considered distributions, we theoretically study the behavior of R as a function of σ_0 by properly fixing the parameters appearing in Equation (4). In particular, we set $\langle m \rangle = 2.23$, $\eta = 0.145$ and consider two possible values of m_μ, 0.1 and 0.04. The resulting expressions for R are plotted in Figure 6.

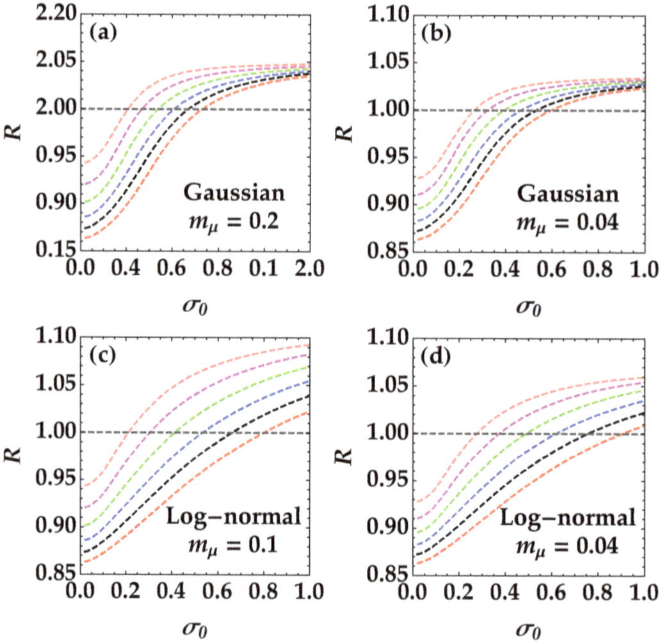

Figure 6. Theoretical values of the noise reduction factor as a function of the standard deviation σ_0 of the Gaussian (panels (**a**,**b**)) and log-normal (panels (**c**,**d**)) distributions for different values of t_0. For all the shown curves we set $\langle m \rangle = 2.23$, $\eta = 0.145$. In panels (**a**,**c**) $m_\mu = 0.1$, while in panel (**b**,**d**) $m_\mu = 0.04$. In each panel, the pink curve corresponds to $t_0 = 0.4$, the magenta curve to $t_0 = 0.5$, the green curve to $t_0 = 0.6$, the blue curve to $t_0 = 0.7$, the black curve to $t_0 = 0.8$ and the red curve to $t_0 = 0.9$.

We note that the choice $\langle m \rangle = 2.23$, and $m_\mu = 0.1$ corresponds to a twin-beam state with $\sim$22 modes, while $\langle m \rangle = 2.23$, and $m_\mu = 0.04$ corresponds to a twin-beam state with $\sim$56 modes, which represents a more reliable experimental condition than the first choice.

The curves shown in panels (a) and (b) are for Gaussian distributions, while those in panels (c) and (d) are for log-normal distributions. In general, the plots resemble the experimental behavior of the plots in Figure 5. However, we notice that depending on the choice of m_μ, the value of σ_0 at which the noise reduction factor is equal to 1, namely the boundary between classical and nonclassical correlations, changes. In particular, the lower the value of m_μ, the larger the threshold value of σ_0. Indeed, according to Equation (4), the lower the value of m_μ, the lower the value of R. To better investigate this result, in Figure 7 we plot the values of σ_0 at which, for fixed choices of $\langle m \rangle$ and η and for 4 possible choices of m_μ, the theoretical value of R is equal to 1 as a function of t_0. We show the results in the case of Gaussian distributions in panel (a) and those for log-normal distributions in panel (b). In general, we can see that the threshold values of σ_0 increase at increasing values of t_0 with different slopes for the two distributions. Moreover, for log-normal distributions of t the growth is more rapid than for Gaussian distributions.

The direct comparison among the curves corresponding to the same kind of distribution leads us to conclude that reducing the mean number of photons per mode, that is increasing the number of modes, makes it possible the detection of nonclassical correlations in the case of wider distributions of t, that is for a highly fluctuating transmittance of the communication channel. Thus, a proper tailoring of the mode structures of the state can make the employed twin beam more robust to losses. We also notice that all the results achieved so far encourage us to test our optical states in more realistic situations. Indeed, as reported in Ref. [20], the probability distributions of t corresponding to free-space quantum channels under diverse weather conditions are characterized by mean values that range from 0.9 down to 0.3. In some cases, the model that describes the transmission efficiency is more symmetric thus resembling a Gaussian distribution, whereas in other situations is more similar to a log-normal distribution.

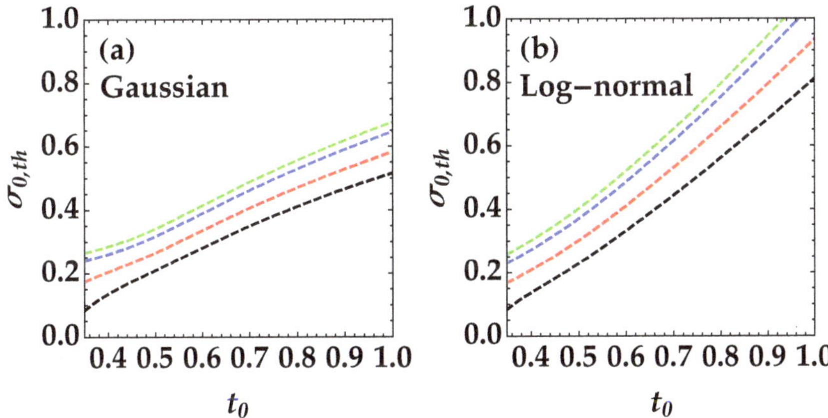

Figure 7. Expected threshold value of σ_0 obtained by setting $R = 1$ in Equation (4) for $\langle m \rangle = 2.23$ and $\eta = 0.145$. The data are shown as a function of t_0 for different choices of m_μ: 0.2 (black curve), 0.1 (red curve), 0.04 (blue curve) and 0.02 (green curve). Panel (**a**) corresponds to the case of Gaussian distributions of t_0, while panel (**b**) to the case of log-normal distributions.

5. Conclusions

In this paper we explored the survival of nonclassical correlations between the parties of mesoscopic twin-beam states propagating through an asymmetric lossy channel, in which the transmission coefficient was statistically distributed. In particular, we considered the case of Gaussian distributions and that of log-normal ones. We investigated the role played by the parameters characterizing the light (mean number of photons and mean number of photons per mode), the transmission channel (mean value and standard deviation of the distribution) and the detection chain (quantum efficiency). In general, we can conclude, as expected, that the larger the mean value of the distribution the better the observation of nonclassicality. Moreover, at a given mean value of the distribution, the larger its standard deviation the worse the observation of nonclassicality. In general, these results hold for both the Gaussian and the log-normal distributions. Not surprisingly, the different shape of the distributions is responsible for the different behavior for the same choice of parameters. In addition, we studied the dependence of the robustness of sub-shot-noise correlations undergoing a statistically distributed amount of loss on the number of modes. We found that high multimode twin-beam states are more robust to loss than low-multimode ones. This result can be ascribed to the fact that a twin beam endowed with many modes has a photon-number distribution less thermal, and thus less fluctuating, than a single-mode one [36–38]. Such a result suggests that, in order to optimize the propagation of light in the presence of loss, a proper tailoring of the employed quantum state should be performed in advance.

Author Contributions: Both authors equally contributed to each step of the work. All authors have read and agreed to the published version of the manuscript.

Funding: This research received no external funding.

Acknowledgments: A. A. acknowledges the Project "Investigating the effect of noise sources in the free-space transmission of mesoscopic quantum states of light" supported by the University of Insubria.

Conflicts of Interest: The authors declare no conflict of interest.

Abbreviations

The following abbreviations are used in this manuscript:

HWP   Half-wave plate
PBS   Polarizing cube beam splitter

References

1. Bennett, C.H.; Bessette, F.; Brassard, G.; Salvail, L.; Smolin, J. Experimental Quantum Cryptography. *J. Cryptol.* **1992**, *5*, 3–28. [CrossRef]
2. Muller, A.; Zbinden, H.; Gisin, N. Underwater quantum coding. *Nature* **1995**, *378*, 449. [CrossRef]
3. Marcikic, I.; de Riedmatten, H.; Tittel, W.; Zbinden, H.; Gisin, N. Long-distance teleportation of qubits at telecommunication wavelengths. *Nature* **2003**, *421*, 509–513. [CrossRef] [PubMed]
4. Ursin, R.; Jennewein, T.; Aspelmeyer, M.; Kaltenbaek, R.; Lindenthal, M.; Walther, P.; Zeilinger, A. Communications: Quantum teleportation across the Danube. *Nature* **2004**, *430*, 849. [PubMed]
5. Kurtsiefer, C.; Zarda, P.; Halder, M.; Weinfurter, H.; Gorman, P.M.; Tapster, P.R.; Rarity, J.G. A step towards global key distribution. *Nature* **2002**, *419*, 450. [CrossRef] [PubMed]
6. Aspelmeyer, M.; Böhm, H.R.; Gyatso, T.; Jennewein, T.; Kaltenbaek, R.; Lindenthal, M.; Molina-Terriza, G.; Poppe, A.; Resch, K.; Taraba, M.; et al. Long-Distance Free-Space Distribution of Quantum Entanglement. *Science* **2003**, *301*, 621. [CrossRef]
7. Jin, X.M.; Ren, J.G.; Yang, B.; Yi, Z.H.; Zhou, F.; Xu, X.F.; Wang, S.K.; Yang, D.; Hu, Y.F.; Jiang, S.; et al. Experimental free-space quantum teleportation. *Nat. Photon.* **2010**, *4*, 376–381. [CrossRef]
8. Capraro, I.; Tomaello, A.; Dall'Arche, A.; Gerlin, F.; Ursin, R.; Vallone, G.; Villoresi, P. Impact of turbulence in long range quantum and classical communications. *Phys. Rev. Lett.* **2012**, *109*, 200502.
9. Peuntinger, C.; Heim, B.; Müller, C.R.; Gabriel, C.; Marquardt, C.; Leuchs, G. Distribution of squeezed states through an atmospheric channel. *Phys. Rev. Lett.* **2014**, *113*, 060502. [CrossRef]
10. Vallone, G.; Bacco, D.; Dequal, D.; Gaiarin, S.; Luceri, V.; Bianco, G.; Villoresi, P. Experimental Satellite Quantum Communications. *Phys. Rev. Lett.* **2015**, *115*, 040502. [CrossRef]

11. Carrasco-Casado, A.; Kunimori, H.; Takenaka, H.; Kubo-Oka, T.; Akioka, M.; Fuse, T.; Koyama, Y.; Kolev, D.; Munemasa, Y.; Toyoshima, M. LEO-to-ground polarization measurements aiming for space QKD using Small Optical TrAnsponder (SOTA). *Opt. Express* **2016**, *24*, 12254. [CrossRef] [PubMed]

12. Dequal, D.; Vidarte, L.T.; Rodriguez, V.R.; Vallone, G.; Villoresi, P.; Leverrier, A.; Diamanti, E. Feasibility of satellite-to-ground continuous-variable quantum key distribution. *arXiv* **2020**, arXiv:2002.02002.

13. Semenov, A.A.; Vogel, W. Quantum light in the turbulent atmosphere. *Phys. Rev. A* **2009**, *80*, 021802(R). [CrossRef]

14. Usenko, V.C.; Heim, B.; Peuntinger, C.; Wittmann, C.; Marquardt, C.; Leuchs, G.; Filip, R. Entanglement of Gaussian states and the applicability to quantum key distribution over fading channels. *New J. Phys.* **2012**, *14*, 093048. [CrossRef]

15. Bohmann, M.; Kruse, R.; Sperling, J.; Silberhorn, C.; Vogel, W. Probing free-space quantum channels with laboratory-based experiments. *Phys. Rev. A* **2017**, *95*, 063801. [CrossRef]

16. Michálek, V.; Perina, J., Jr.; Haderka, O. Experimental Quantification of the Entanglement of Noisy Twin Beams. *Phys. Rev. Appl.* **2020**, *14*, 024003.

17. Allevi, A.; Bondani, M. Preserving nonclassical correlations in strongly unbalanced conditions. *J. Opt. Soc. Am. B* **2019**, *36*, 3275–3281. [CrossRef]

18. Straka, I.; Mika, J.; Ježek, M. Generator of arbitrary classical photon statistics. *Opt. Express* **2018**, *26*, 8998–9010. [CrossRef]

19. Milonni, P.W.; Carter, J.H.; Peterson, C.G.; Hughes, R.J. Effects of propagation through atmospheric turbulence on photon statistics. *J. Opt. B Quantum Semiclass. Opt.* **2004**, *6*, S742–S745. [CrossRef]

20. Vasylyev, D.; Semenov, A.A.; Vogel, W.; Günthner, K.; Thurn, A.; Bayraktar, Ö.; Marquardt, C. Free-space quantum links under diverse weather conditions. *Phys. Rev. A* **2017**, *96*, 043856.

21. Agliati, A.; Bondani, M.; Andreoni, A.; De Cillis, G.; Paris M.G.A. Quantum and classical correlations of intense beams of light via joint photodetection. *J. Opt. B* **2005**, *7*, 652 . [CrossRef]

22. Lee, C.T. Many-photon antibunching in generalized pair coherent states. *Phys. Rev. A* **1990**, *41*, 1569. [CrossRef] [PubMed]

23. Lee, C.T. Nonclassical photon statistics of two-mode squeezed states. *Phys. Rev. A* **1990**, *42*, 1608. [CrossRef] [PubMed]

24. Peres, A. Separability criterion for density matrices. *Phys. Rev. Lett.* **1996**, *77*, 1413. [CrossRef]

25. Simon, R. Peres-Horodecki Separability Criterion for Continuous Variable Systems. *Phys. Rev. Lett.* **2000**, *84*, 2726. [CrossRef]

26. Richter, T.; Vogel, W. Nonclassicality of quantum states: A hierarchy of observable conditions. *Phys. Rev. Lett.* **2002**, *89*, 283601. [CrossRef]

27. Degiovanni, I.P.; Genovese, M.; Schettini, V.; Bondani, M.; Andreoni, A.; Paris, M.G.A. Monitoring the quantum-classical transition in thermally seeded parametric down-conversion by intensity measurements. *Phys. Rev. A* **2009**, *79*, 063836. [CrossRef]

28. Allevi, A.; Olivares, S.; Bondani, M. Measuring high-order photon-number correlations in experiments with multimode pulsed quantum states. *Phys. Rev. A* **2012**, *85*, 063835. [CrossRef]

29. Brida, G.; Genovese, M.; Ruo Berchera, I. Experimental realization of sub-shot-noise quantum imaging. *Nat. Photon.* **2010**, *4*, 227–230.

30. Chan, K.W.; O'Sullivan, M.N.; Boyd, R.W. High-order thermal ghost imaging. *Opt. Lett.* **2009**, *34*, 3343–3345. [CrossRef]

31. Bondani, M.; Allevi, A.; Andreoni, A. Ghost imaging by intense multimode twin beam. *Eur. Phys. J. Spec. Top.* **2012**, *203*, 151–161. [CrossRef]

32. Allevi, A.; Bondani, M. Can nonclassical correlations survive in the presence of asymmetric lossy channels? *Eur. Phys. J. D* **2018**, *72*, 178. [CrossRef]

33. Allevi, A.; Bondani, M. Nonlinear and Quantum Optical Properties and Applications of Intense Twin-Beams. *Adv. At. Mol. Opt. Phys.* **2017**, *66*, 49–110.

34. Bondani, M.; Allevi, A.; Agliati, A.; Andreoni, A. Self-consistent characterization of light statistics. *J. Mod. Opt.* **2009**, *56*, 226–231. [CrossRef]

35. Allevi, A.; Lamperti, M.; Bondani, M.; Peřina, J., Jr.; Michálek, V.; Haderka, O.; Machulka, R. Characterizing the nonclassicality of mesoscopic optical twin-beam states. *Phys. Rev. A* **2013**, *88*, 063807. [CrossRef]

36. Lamperti, M.; Allevi, A.; Bondani, M.; Machulka, R.; Michálek, V.; Haderka, O.; Peřina, J., Jr. Optimal sub-Poissonian light generation from twin beams by photon-number resolving detectors. *J. Opt. Soc. Am. B* **2014**, *31*, 20–25.
37. Allevi, A.; Bondani, M. Statistics of twin-beam states by photon-number resolving detectors up to pump depletion. *J. Opt. Soc. Am. B* **2014**, *31*, B14–B19. [CrossRef]
38. Arkhipov, I.I.; Peřina, J., Jr.; Peřina, J.; Miranowicz, A. Comparative study of nonclassicality, entanglement, and dimensionality of multimode noisy twin beams. *Phys. Rev. A* **2015**, *91*, 033837. [CrossRef]

Publisher's Note: MDPI stays neutral with regard to jurisdictional claims in published maps and institutional affiliations.

 *applied sciences*

Article

Quantum Photonic Simulation of Spin-Magnetic Field Coupling and Atom-Optical Field Interaction

Jesús Liñares *,†, Xesús Prieto-Blanco †, Gabriel M. Carral † and María C. Nistal †

Quantum Materials and Photonics Research Group, Optics Area, Department of Applied Physics,
Faculty of Physics/Faculty of Optics and Optometry, Campus Vida s/n, University of Santiago de Compostela,
E-15782 Santiago de Compostela, Galicia, Spain; xesus.prieto.blanco@usc.es (X.P.-B.);
gabrielmaria.carral@rai.usc.es (G.M.C.); mconcepcion.nistal@usc.es (M.C.N.)
* Correspondence: suso.linares.beiras@usc.es
† These authors contributed equally to this work.

Received: 8 November 2020; Accepted: 7 December 2020; Published: 10 December 2020

Abstract: In this work, we present the physical simulation of the dynamical and topological properties of atom-field quantum interacting systems by means of integrated quantum photonic devices. In particular, we simulate mechanical systems used, for example, for quantum processing and requiring a very complex technology such as a spin-1/2 particle interacting with an external classical time-dependent magnetic field and a two-level atom under the action of an external classical time-dependent electric (optical) field (light-matter interaction). The photonic device consists of integrated optical waveguides supporting two collinear or codirectional modes, which are coupled by integrated optical gratings. We show that the single-photon quantum description of the dynamics of this photonic device is a quantum physical simulation of both aforementioned interacting systems. The two-mode photonic device with a single-photon quantum state represents the quantum system, and the optical grating corresponds to an external field. Likewise, we also present the generation of Aharonov–Anandan geometric phases within this photonic device, which also appear in the simulated systems. On the other hand, this photonic simulator can be regarded as a basic brick for constructing more complex photonic simulators. We present a few examples where optical gratings interacting with several collinear and/or codirectional modes are used in order to illustrate the new possibilities for quantum simulation.

Keywords: integrated photonics; quantum optics; quantum simulation

---

1. Introduction

One of the most promising tasks in quantum science and technology is the implementation of quantum simulations. Its physical foundation is based on the fact that the dynamics of a quantum system is governed by its Hamiltonian $\hat{H}$ (time evolution) or momentum operator $\hat{M}$ (spatial evolution), that is given a Hilbert space $\mathcal{H}$ and some input state $|\Psi(0)\rangle \in \mathcal{H}$, the full evolution of the system is given by the action of the evolution operator on such a state. The evolution operator can be either the time evolution one $\hat{U}_t = \exp(-i\hat{H}t/\hbar)$, which comes from the Schrödinger equation $-i\hbar\partial|\Psi\rangle/\partial t = \hat{H}|\Psi\rangle$, that is $|\Psi(t)\rangle = \hat{U}_t|\Psi(0)\rangle$, or its spatial counterpart $\hat{U}_s = \exp(i\hat{M}s/\hbar)$, which comes from the momentum operator in the position representation, that is $i\hbar\partial|\Psi\rangle/\partial s = \hat{M}|\Psi\rangle$, where s is the spatial variable that defines the direction along which the system evolves (it can be, for instance, the z-direction) then $|\Psi(s)\rangle = \hat{U}_s|\Psi(0)\rangle$. One is perhaps more familiar with the temporal case $|\Psi(t)\rangle = \hat{U}_t|\Psi(0)\rangle$, but note that the spatial case is totally analogous [1,2]. Now, as it occurs in nature that very different and unrelated quantum systems share analogous Hamiltonians or momentum operators, the dynamics of these systems will be analogous. This implies that if we have some system of interest, we will be able to mimic its behavior (evolution) by means of other very different system. This last system, which is required to be fully controllable by the experimenter, constitutes a quantum simulator [3,4] of the system of interest.

On the other hand, the importance of making quantum simulations obeys various reasons. First of all, whether classical or quantum, simulation is often an indispensable tool in science. Physical systems can be very complicated to study. If we find a way to mimic their behavior in a manageable way, their dynamics can be analyzed with much less difficulty, much less expensively, and much faster [3]. For instance, we can imagine that the system of interest is one whose working conditions are fragile and very sensitive to small perturbations. It could also be that those conditions are very specific and/or hard (even impossible) to achieve in the laboratory. Simulating these complicated systems will, in principle, improve our knowledge about them with much ease. Secondly, quantum simulation offers a crucial advantage over classical simulation. Quantum systems have a greater information storing capacity than classical systems; thus, a quantum simulator, which is a quantum system itself, will require much fewer physical resources than a classical simulator to store the same amount of information. Furthermore, classical simulations find difficulties simulating quantum systems when strong entanglement is present [3]. Finally, it is possible that the potential applications of a quantum system may be better implemented using a quantum simulation [5], if it happens that the quantum simulation parameters are easier to handle, compared to those of the quantum system of interest, in the sense of greater tunability [4]. All of this strongly suggests the need for quantum simulations. Formally, one has a quantum device, called the quantum simulator, which reproduces the evolution of the system of interest, the quantum system. Following [4], let us call the initial input state of the quantum system $|\phi(0)\rangle$, which evolves to $|\phi(\zeta)\rangle$, where ζ stands either for time or some spatial coordinate, under the evolution operator $\hat{U} = \exp(i\hat{O}\zeta/\hbar)$. Here, $\hat{O}$ can be a Hamiltonian $(-\hat{H})$ or a momentum operator $(\hat{M})$. The quantum simulator, on the other hand, starts in the state $|\psi(0)\rangle$ and evolves to $|\psi(\zeta)\rangle$ under the action of the operator $\hat{U}' = \exp(i\hat{O}'\zeta/\hbar)$. Since we have a system and a simulator, there exists a correspondence relating these elements, that is a correspondence between $|\phi(0)\rangle \leftrightarrow |\psi(0)\rangle$, $|\phi(\zeta)\rangle \leftrightarrow |\psi(\zeta)\rangle$, $\hat{U} \leftrightarrow \hat{U}'$, and thus, $\hat{O} \leftrightarrow \hat{O}'$. This is revealed via measurements on both systems. The greater the accuracy of this correspondence, the more one can trust the simulator [3].

In this work, we present an integrated quantum photonic simulator for atom-field quantum interacting systems. It is based on optical gratings and can be regarded as a basic brick for constructing more complex photonic simulators. First of all, we would like to stress the high interest in quantum photonic simulators; for instance, integrated photonic structures allow simulating a number of condensed matter effects such as Anderson localization, Mott transition, etc. [6], and more recently, anyonic interaction has been simulated by using an array of channel optical waveguides with a helically bent axis [7]. In our case, we will simulate in an integrated photonic device the dynamical and topological properties of two very well-known quantum interacting systems. This is relevant since a way to check the fidelity of a simulation is testing the same physical model of the dynamics with different systems and then comparing the results [5]. The two systems to be simulated are the well-known interaction of a spin-1/2 particle under a time-dependent external classical magnetic field (see for instance [8]) and the interaction of a two-level atom with an external classical electric field (see for instance [9]) which are used, at present, for implementing quantum information devices for quantum processing, quantum computation, and so on [8]. It is also well known that these mechanical systems require a very complex technology. We will take a standard photonic device, that is a two-mode planar guide modulated by an integrated grating (see for instance [10]) in which a single-photon quantum state is propagated. We will show that it can emulate both the dynamical and geometrical properties of the two aforementioned systems. As mentioned, this photonic simulator based on integrated optical gratings can be regarded as a basic brick for more complex photonic simulators; thus, the results obtained can be extended to multi-level systems or to several concatenated two-mode systems simulating concatenated temporal operations. Addressing practical issues, this simulator will also benefit from some advantages of classical and quantum integrated photonics [6,11,12], such as the fast and energetically efficient operation and miniaturization capabilities of integrated photonics, which favor scalability. Moreover, integrated photonic devices are becoming some of the most powerful and appealing technologies for classical and quantum information [13,14]. Finally, we must stress that although we present a photonic device for quantum simulation, it can be also used to implement quantum operations and/or effects with the photonic device itself such as logic gates, geometric phases, and so on.

The plan of the paper is as follows: In Section 2, we briefly present the Hamiltonian of the quantum interacting systems, in a suitable form for their photonic simulation, along with the physical parameters and some useful solutions. In Section 3, starting from quantum states as single-photon states, an integrated photonic device for the quantum simulation of both a spin-magnetic field coupling and light-matter interaction is presented, along with the limits of this simulation. In Section 4, quantum simulation of geometric phases is studied. Finally, a summary is presented in Section 5.

2. Mechanical Interacting Systems

In this section, we present the main results about the mechanical interacting systems whose dynamic and topological properties are going to be simulated with an integrated photonic device. The primary aim is to write the Hamiltonians of these systems in a suitable form to facilitate the study of their photonic simulations.

2.1. Spin-1/2 Particle Interacting with a Magnetic Field

The Hamiltonian of a spin particle interacting with a magnetic field is given by $\hat{H} = -\boldsymbol{\mu} \cdot \boldsymbol{B}$, where $\boldsymbol{\mu}$ is the magnetic moment of the particle, which is proportional to the spin operator, and $\boldsymbol{B}$ is the magnetic field. Let us consider a spin-$\frac{1}{2}$ particle, then $\boldsymbol{\mu} = \frac{1}{2}\mu\boldsymbol{\sigma}$, where $\mu = \hbar\gamma_p$, with γ_p the gyromagnetic ratio of the particle ($p =$ e for an electron, $p =$ n for a neutron, etc.), and $\boldsymbol{\sigma} = (\sigma_x, \sigma_y, \sigma_z)$ are the Pauli matrices. Moreover, we choose a time-dependent magnetic field, which, by assuming a sinusoidal dependence in time with frequency ω, is written as $\boldsymbol{B} = B_0(\sin\theta\cos\omega t, -\sin\theta\sin\omega t, \cos\theta)$, representing a magnetic field of modulus B_0 rotated an angle θ with respect to the z-axis and spinning around this same axis with a frequency ω. We can then write the well-known spin-magnetic Hamiltonian as (see for instance [8]):

$$\hat{H}(t) = -\frac{\hbar}{2}\gamma_p B_0 \cos\theta\sigma_z - \frac{\hbar}{2}\gamma_p B_0 \sin\theta(\cos\omega t\,\sigma_x - \sin\omega t\,\sigma_y). \tag{1}$$

By taking into account the expressions of the Pauli matrices, then the above Hamiltonian gives rise to the following time-dependent Schrödinger equation:

$$i\hbar\frac{\partial}{\partial t}|\Psi(t)\rangle = -\frac{\hbar}{2}\gamma_p B_0 \begin{pmatrix} \cos\theta & \sin\theta\exp(i\omega t) \\ \sin\theta\exp(-i\omega t) & -\cos\theta \end{pmatrix}|\Psi(t)\rangle =$$

$$\equiv \hat{H}(t)|\Psi(t)\rangle. \tag{2}$$

This time-dependent Hamiltonian is enough for our simulation purposes. Note that for the neutron case, we have the well-known Nuclear Magnetic Resonance (NMR). The solution of the equation above can be obtained as a time-dependent linear combination of down and up spin states, $|0\rangle$ and $|1\rangle$, of the particle, that is $|\Psi(t)\rangle = c_0(t)|0\rangle + c_1(t)|1\rangle$. Thus, by using the vector representation of $|\Psi(t)\rangle$ and the matrix Hamiltonian given by Equation (2), the Schrödinger equation can be rewritten as follows:

$$i\hbar\frac{dc_n(t)}{dt} = E_n\cos\theta c_n(t) + i\sum_{n\neq m}C_{nm}(t)c_m(t). \tag{3}$$

with $n = 0, 1$, $E_o = -\frac{\hbar \gamma_p}{2} B_0$ and $E_1 = \frac{\hbar \gamma_p}{2} B_0$ the eigenvalues of spin states when a static magnetic field along z-direction is applied, and $C_{01}(t) = C_{10}^*(t) = -\frac{\hbar \gamma_p}{2} B_0 \sin \theta \exp(i\omega t)$ the coupling coefficients. On the other hand, it is interesting, for the sake of expositional convenience, to present the main results about the dynamics of the system. First of all, in order to simplify things, we can shift to a reference frame that is rotating at a frequency ω by means of the following unitary transformation:

$$|\eta(t)\rangle = \hat{U}(t, \omega)|\Psi(t)\rangle = \exp\left(-i\omega\sigma_z t/2\right)|\Psi(t)\rangle; \tag{4}$$

therefore, by inserting $|\Psi(t)\rangle = \exp\left(i\omega\sigma_z t/2\right)|\eta(t)\rangle$ into the Schrödinger equation, we obtain, after some simple calculations, the following Hamiltonian acting on $|\eta(t)\rangle$:

$$\hat{H}'|\eta(t)\rangle = \frac{\hbar}{2} \begin{pmatrix} -\gamma_p B_0 \cos \theta + \omega & -\gamma_p B_0 \sin \theta \\ -\gamma_p B_0 \sin \theta & \gamma_p B_0 \cos \theta - \omega \end{pmatrix} |\eta(t)\rangle. \tag{5}$$

This Hamiltonian $\hat{H}'$ can be rewritten as a product of the form $-\frac{\hbar}{2}\gamma_p(\sigma \cdot \boldsymbol{B}_{ef})$, with $\boldsymbol{B}_{ef}$ an effective static magnetic field, that is,

$$\hat{H}' = -\frac{\hbar}{2}\gamma_p(\sigma \cdot \boldsymbol{B}_{ef}) = -\frac{\hbar}{2}\gamma_p B_0 \Delta \cos \theta' \sigma_z - \frac{\hbar}{2}\gamma_p B_0 \Delta \sin \theta' \sigma_x = -\frac{\hbar}{2}\Delta_o \begin{pmatrix} \cos \theta' & \sin \theta' \\ \sin \theta' & -\cos \theta' \end{pmatrix}, \tag{6}$$

with:

$$\Delta_o = \gamma_p B_0 \Delta = \gamma_p B_o' \tag{7}$$

and:

$$\Delta = \gamma_p B_0 \sqrt{1 - \frac{2\omega}{\gamma_p B_0} \cos \theta + \frac{\omega^2}{\gamma_p^2 B_0^2}}; \tag{8}$$

therefore, the effective magnetic field is given by $\boldsymbol{B}_{ef} = (B_0' \sin \theta', 0, B_0' \cos \theta')$, where B_0' and θ' are related to B_0 and θ by $B_0' = B_0 \Delta$, with $\sin \theta' = \gamma_p B_0 \sin \theta / \Delta_o = \sin \theta / \Delta$ and $\cos \theta' = \gamma_p B_0 [\cos \theta - (\omega/\gamma_p B_0)]/\Delta_o = [\cos \theta - (\omega/\gamma_p B_0)]/\Delta$. As seen from the expression for Δ, it would seem that for this reparametrization to have physical meaning, restrictions on the parameters would appear, as Δ contains a possible non-positive term under the square root. The worst case would be likely to happen if $\cos \theta = 1$, but one can find that the resulting quantity $1 - (2\hbar\omega/\mu B_0) + (\hbar\omega/\mu B_0)^2$ is never negative, for any value of the parameters.

Next, it is interesting to compute the eigenstates $|\eta\rangle$ and eigenvalues E_η of Hamiltonian $\hat{H}'$ given by Equation (6), that is $|\eta(t)\rangle = \exp\left(-\frac{i}{\hbar}E_\eta t\right)|\eta(0)\rangle$. After a standard calculation, we have:

$$|\eta_+(0)\rangle = \cos\frac{\theta'}{2}|0\rangle + \sin\frac{\theta'}{2}|1\rangle, \quad |\eta_-(0)\rangle = \sin\frac{\theta'}{2}|0\rangle - \cos\frac{\theta'}{2}|1\rangle, \tag{9}$$

where $|0\rangle \equiv (1,0)^T$ and $|1\rangle \equiv (0,1)^T$, with T denoting the transpose. The eigenvalues are $E_\pm = \pm\frac{\hbar}{2}\Delta_o$, that is $E_\pm = \pm\frac{\hbar}{2}\gamma_p B_0' = \pm\frac{\mu}{2}B_0'$. Therefore, by taking into account Equation (4), the full evolved states $|\Psi(t)\rangle$ can be easily obtained.

2.2. Two-Level Atom Interacting with an Electric (Optical) Field

On the other hand, let us consider a two-level atom coupled to a harmonic external classical electric field, which describes semiclassical light-matter interaction. As is usually done [9], we label the atomic levels as $|g\rangle$ (ground) and $|e\rangle$ (excited). They have energies $\hbar\omega_g$ and $\hbar\omega_e$, respectively.

Their energy difference is given by $\hbar w_0 = \hbar\,(w_e - w_g)$. The expression for this non-interacting part of the Hamiltonian can be formally written as follows:

$$\hat{H}_o = \hbar w_g |g\rangle\langle g| + \hbar w_e |e\rangle\langle e| = \hbar\bar{w}\mathbf{I} + \hbar w_o \sigma_z/2, \tag{10}$$

where we used the matrix representation of $|g\rangle\langle g|$ and $|e\rangle\langle e|$, with $\bar{w} = (w_g + w_e)/2$, and $\mathbf{I}$ the two-dimensional identity matrix. As for the interacting part, it is given by the dipole interaction (electric dipole approximation) between an external electric (optical) field and the atom. Indeed, by assuming, for the sake of simplicity, that the field is propagating along z, has a sinusoidal dependence in time with frequency w, and is linearly polarized along the x-direction, then the electric (optical) field can be written as follows $E = E_0 \cos\left(w(z/c - t)\,\mathbf{u}_x\right.$. Moreover, we disregard the spatial dependence of the electric (optical) field because the wavelength $\lambda = c/2\pi w$ is considered much larger than the atomic dimensions. Therefore, we apply the dipole or long-wavelength approximation, that is by assuming without lost of generality $wz/c = 2m\pi$, with m an integer, then $E = E_0 \cos wt\,\mathbf{u}_x$, the electric-dipole interaction is given by $\hat{H}_I = -\mathbf{d}\cdot\mathbf{E} = -d_x E_x = exE_x$, where $\mathbf{d}$ is the atomic dipole operator $\mathbf{d} = q\cdot\mathbf{r}$, with $q = -e$, and $\mathbf{r}$ is the electron's position vector (operator). This interaction term allows transitions between the two levels. The form of the dipole operator can be calculated by using twice the closure relationship $\hat{I} = |g\rangle\langle g| + |e\rangle\langle e|$, that is $\hat{I}(-ex)\hat{I}$; therefore:

$$d_x = -\langle g|ex|g\rangle|g\rangle\langle g| - \langle e|ex|e\rangle|e\rangle\langle e| - \langle g|ex|e\rangle(|g\rangle\langle e| + |e\rangle\langle g|) \equiv -c_+\mathbf{I} + c_-\sigma_z - d_o\sigma_x, \tag{11}$$

with $c_\pm = (\langle g|ex|g\rangle \pm \langle e|ex|e\rangle)/2$, $d_0 = \langle g|ex|e\rangle$, and where we have used the matrix representations $(|g\rangle\langle e| + |e\rangle\langle g| \equiv \sigma_x$, $|g\rangle\langle g| \equiv (\mathbf{I} + \sigma_z)/2$, and $|e\rangle\langle e| = (\mathbf{I} - \sigma_z)/2$. We must stress that sometimes, parity arguments [9] narrow the form of the dipole operator, leading up to the expression $d_x = -\langle g|ex|e\rangle(|g\rangle\langle e| + |e\rangle\langle g|) = -d_o\sigma_x$. Likewise, it is customary to introduce the Rabi frequency $\Omega = E_o d_o/\hbar$. In short, the total Hamiltonian is given by $\hat{H}_o + \hat{H}_I$, and therefore, the corresponding Schrödinger equation, by using this new variables, is given by:

$$i\hbar\frac{\partial}{\partial t}|\Psi(t)\rangle = \left(\hbar(\bar{w} + C_+)\,\mathbf{I} - \hbar(\frac{w_0}{2} - C_-)\sigma_z + \hbar\Omega\sigma_x \cos wt\right)|\psi(t)\rangle = \hat{H}|\Psi(t)\rangle, \tag{12}$$

with $C_\pm = c_\pm \cos wt$. The general solution of this Schrödinger equation can again be obtained by using a time-dependent linear combination of fundamental and excited states, $|g\rangle \equiv |0\rangle$ and $|e\rangle \equiv |1\rangle$, of the particle, that is $|\Psi(t)\rangle = c_0(t)|0\rangle + c_1(t)|1\rangle$. However, the high value of the frequency w means that the electric field is rapidly oscillating, which suggests to make the following change $|\Psi(t)\rangle = f_0(t)\exp(iwt/2)|0\rangle + f_1(t)\exp(-iwt/2)|1\rangle \equiv \exp(iw\sigma_z t/2)|\eta(t)\rangle$. Moreover, in many cases, by parity arguments, it is fulfilled that $c_\pm = 0$. Therefore, by substituting this state into Equation (12), we obtain the following Hamiltonian for the state $(f_0(t), f_1(t))^T \equiv |\eta(t)\rangle$,

$$\hat{H}'|\eta(t)\rangle \approx \left(\hbar\bar{w}\,\mathbf{I} - \frac{\hbar\delta}{2}\sigma_z + \frac{\hbar\Omega}{2}\sigma_x\right)|\eta(t)\rangle, \tag{13}$$

with $\delta = (w_0 - w)$ the detuning parameter and where we have neglected terms rapidly oscillating of the form $\exp(\pm iwt)$; that is, a temporal Rotating Wave Approximation (RWA) has been made, and a time independent Hamiltonian has thus been obtained. We must stress that under this approximation, the terms $C_\pm$ in Equation (12) can be also neglected independently of the wave-function parity. Finally, note that Hamiltonians given by Equations (6) and (13) have the same algebraic structure.

3. Quantum Photonic Simulations

In this section, we present the quantum simulation, which can be implemented by an integrated photonic device, that is integrated optical gratings supporting two collinear guided modes, that is two mode guides assisted by a periodic perturbation. It can be considered the basic brick for constructing more complex simulators.

3.1. Classical Study of the Photonic Device

Let us consider a standard integrated photonic device consisting of, for example, an integrated waveguide 1D (one-dimensional) with refractive index $n(x)$ (slab guide) that supports two optical modes $e_0(x)$ and $e_1(x)$. These modes are collinear and travel in the z-direction with propagation constants β_0 and β_1, that is wave vectors in the z-direction defined as $\beta = (\omega/c)N$, where ω is the frequency of the mode and N is the effective index [10]. The mentioned modes are coupled by an integrated grating present in a region of the slab guide, as shown in Figure 1a), where relevant parameters are indicated, that is substrate index n_s, film index n_f, cover index $n_c = 1$, and film thickness d. The grating is represented, for example, by a periodic modulation (perturbation) of the electrical permittivity [10],

$$\Delta\epsilon(x,z) = \Delta\epsilon(x)\cos(\gamma z + \alpha_0), \tag{14}$$

with $\Delta\epsilon(x)$ the modulation strength of the optical grating, α_o an initial phase, if required, $\gamma = 2\pi/\Lambda$ the frequency of the perturbation, and Λ its period. This index profile can be obtained by different technologies of integrated optics, for instance ion-exchange in glass [15,16] could be used, or even by optical fiber technology [17]. Likewise, in crystals such as lithium niobate, these optical gratings can also be reconfigurable due to acousto-optic or electro-optic effects [18].

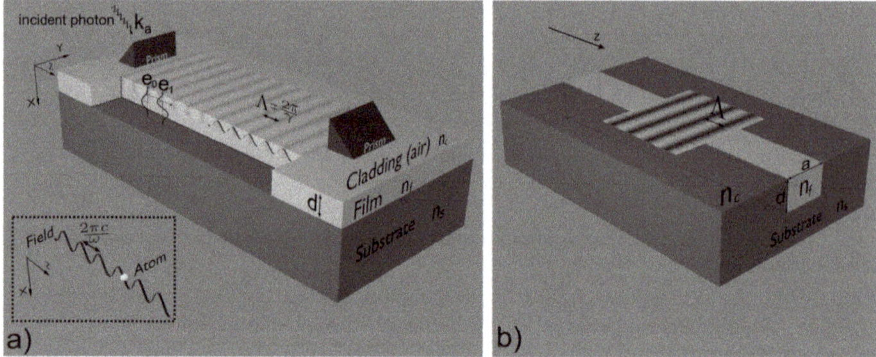

Figure 1. (**a**) Integrated optical grating with period Λ on a two-mode slab waveguide with modes $e_0(x)$ and $e_1(x)$. The inset shows the simulated mechanical device: atom-field interaction. (**b**) Integrated optical grating with period Λ on a two-mode channel waveguide with modes, for example $e_{00}(x,y)$ and $e_{10}(x,y)$.

As we assume that only two guided modes are excited in our integrated photonic structure, the perturbed electric field amplitude is given by $e(x,z) = a_0(z)e_0(x) + a_1(z)e_1(x)$. It is well known that the general set of equations that describe the coupling between n copropagating optical modes with propagation constants β_n in a perturbed waveguide and that allow calculating the amplitude coefficients $a_n(z)$ is given by (see for instance [10]):

$$-i\frac{da_n(z)}{dz} = \tilde{\beta}_n(z)a_n(z) + \sum_{n\neq m} C_{nm}(z)a_m(z), \tag{15}$$

where $\tilde{\beta}_n(z) = \beta_n + C_{nn}(z)$ are the corrected propagation constants due to the self-coupling coefficients C_{nn} and C_{nm} are the coupling coefficients between the n mode and each of the other m modes. All these coefficients are calculated as follows [10]:

$$C_{nm}(z) = \frac{\omega}{2} \int \Delta\epsilon(x,z)\, e_n(x)\, e_m^*(x)\, \mathrm{d}x, \tag{16}$$

with ω the temporal frequency of the modes and $e_n(x)$ and $e_m(x)$ the normalized optical n and m modes of a planar guide.

The study with 2D guides (integrated channel guides or optical fibers) can be also made, but no new relevant result would be obtained. Indeed, a channel guide can be defined starting from a planar guide whose width is reduced up to a size a of the same order as its depth, that is $a \approx d$, as shown in Figure 1b). In such a case, the optical modes are characterized by two subscripts, one for each spatial direction, that is $e_{np}(x,y)$ and $e_{mq}(x,y)$; therefore, the general coupling coefficients are given by:

$$C_{npmq}(z) = \frac{\omega}{2} \int \Delta\epsilon(x,y,z)\, e_{np}(x,y)\, e_{mq}^*(x,y)\, \mathrm{d}x\mathrm{d}y, \tag{17}$$

where we assumed that the grating modulation can also have a y-dependence. Finally, the mode coupling equations can be obtained by applying the formal changes $n \to np$, $m \to mq$ in Equation (15). Accordingly, the results for planar guides can be easily transferred to channel guides.

3.2. Quantum Study of the Photonic Device

A canonical quantization procedure [1,2] proves that the $a(z)$ coefficients become the photon absorption (or emission) operators $\hat{a}(z)$ (correspondence principle). Therefore, the coupled mode equations are in fact the Heisenberg equations, which, in general, give the evolution of the operators in time. In this case, they give the spatial evolution of the operators. Moreover, the relevant operator here responsible for the spatial propagation of quantum states of the device is the momentum operator, which is the generator of spatial translations [2,19], and not the Hamiltonian, which is the generator of temporal translations. In short, we can study the integrated photonic device in a fully quantum mechanical way, by solving the equations for absorption operators (spatial Heisenberg equations). That is, by performing the change $a(z) \to \hbar\hat{a}(z)$ in Equation (15), we obtain [2,20]:

$$-i\hbar\frac{d\hat{a}_n(z)}{dz} = \hbar\tilde{\beta}_n\hat{a}_n(z) + \hbar \sum_{n \neq m} C_{nm}(z)\hat{a}_m(z). \tag{18}$$

We must stress that modal coupling preserves energy; therefore, Equation (18) corresponds to a unitary transformation, and accordingly, $C_{nm} = C_{mn}$. On the other hand, we only consider single-photon states, which is enough for our simulation purposes. We must stress that the linear momentum of a single photon without modal coupling, that is $\Delta\epsilon(x,y) = 0$, is given by $p_{(0,1)} = \hbar\beta_{(0,1)}$ depending on whether the photon is excited in mode β_0 or mode β_1 [19,20]. Obviously, more general quantum states could be used such as multiphoton states, entangled states of two photons, and so on. These states would give rise to more complex quantum simulations, which fall outside of the scope of this work. In general, the single-photon state is a quantum superposition because the photon can either be excited in the mode β_0, that is $|1_0\rangle$, or in the mode β_1, that is $|1_1\rangle$. Hence, the general quantum state is given by:

$$|L(z)\rangle = a_0(z)|1_0\rangle + a_1(z)|1_1\rangle, \tag{19}$$

where $a_0(z)$ and $a_1(z)$ are the quantum complex amplitudes and fulfil the normalization condition $|a_0(z)|^2 + |a_1(z)|^2 = 1$. States $|1_0\rangle$ and $|1_1\rangle$ must be understood as single-photon states at a distance (plane) z. It can be checked that the solutions of Equation (18) for the spatial propagation of emission operators are the same as for single-photon states $|L(z)\rangle$ [21]. The main reason is that single-photon states are proportional to emission operators, that is $|1_0\rangle = \hat{a}_0^\dagger|0\rangle$, $|1_1\rangle = \hat{a}_1^\dagger|0\rangle$. Indeed, let us consider free propagation, that is non-coupling case $C_{nm} = 0$, then the solutions of Equation (18) are $\hat{a}_{0,1}(z) = \exp(i\beta_{0,1}z)\hat{a}_{0,1}(0)$. Now, let us consider, for the sake of simplicity, a single photon state at $z = 0$, for instance $|1_0\rangle$, then the optical propagation can be obtained as follows: $|1_0\rangle = \hat{a}_0^\dagger(0)|0\rangle$ (one-photon emission), but by taking into account the z-propagation, we have $|L(z)\rangle = \exp(i\beta_0 z)\hat{a}_0^\dagger(z)|0\rangle = \exp(i\beta_0 z)|1_0\rangle \equiv a_0(z)|1_0\rangle$; therefore, the coefficients $a_{0,1}(z)$ of a single-photon state have the same optical propagation solution as the operators $\hat{a}_{0,1}(z)$. This can be proven for a general unitary transformation after a certain algebra. We will take advantage of this property for simulation purposes. Accordingly, for single-photon states, Equation (18) can formally be rewritten as follows:

$$i\hbar\frac{d|L(z)\rangle}{dz} = -\hbar \begin{pmatrix} \tilde{\beta}_0(z) & C_{12}(z) \\ C_{21}(z) & \tilde{\beta}_1(z) \end{pmatrix} |L(z)\rangle = -\hat{M}(z)|L(z)\rangle, \tag{20}$$

where $|L(z)\rangle$ is given, in vector representation, by $(a_0(z), a_1(z))^T$ and $\hat{M}$ is the matrix representation of the so-called momentum operator. This is equivalent to the matrix equation given by Equation (2) for the Hamiltonian. We must stress that Equations (2), (3), (13), (18) and (20) are the main results for implementing the quantum simulations in this work.

3.3. Photonic Simulation of Spin-Magnetic Field Interaction

Let us consider an integrated optical grating characterized by the function given by Equation (14). It will be useful to work with the slowly varying operators $\hat{A}_n$, defined as $\hat{A}_n = \hat{a}_n \exp(-i\beta_n z)$. On the other hand, the coupling coefficients for a grating with initial phase $\alpha_o = 0$ can be written as $C_{00} = c_{00}\cos\gamma z$, $C_{11} = c_{11}\cos\gamma z$, and $C_{01} = C_{10} = C_o\cos\gamma z$, where $c_{nm} = (\omega/2)\int \Delta\epsilon(x)e_n(x)\cdot e_m^*(x)dx$, and so on. Therefore, from Equation (18), we obtain, for the two-mode case, the following spatial Heisenberg equations:

$$i\frac{d\hat{A}_0(z)}{dz} = -\hat{A}_0\, c_{00}\cos\gamma z - \hat{A}_1\frac{C_o}{2}e^{-i\Delta\beta_{01}z}[e^{i\gamma z} + e^{-i\gamma z}], \tag{21}$$

$$i\frac{d\hat{A}_1(z)}{dz} = -\hat{A}_1\, c_{11}\cos\gamma z - \hat{A}_0\frac{C_o}{2}e^{i\Delta\beta_{01}z}[e^{i\gamma z} + e^{-i\gamma z}], \tag{22}$$

where $\Delta\beta_{01} = \beta_0 - \beta_1$. The above equation reveals that we have oscillating terms coming from the cosine of the self-coupling terms with arguments $\pm\gamma z$ and also terms with arguments $(\gamma - \Delta\beta_{01})z$ and $(\gamma + \Delta\beta_{01})z$. We make the assumption that γ is of the same order as $\Delta\beta_{01}$, so that $(\gamma - \Delta\beta_{01})$ is small. The other terms are rapidly oscillating and, thus, will average to zero on a sufficiently large z-scale. We must stress that what we do here is essentially a spatial RWA, which is well-known in light-matter interaction in the time domain. Therefore, the above equations become, to a good approximation,

$$i\frac{d\hat{A}_0(z)}{dz} = -\hat{A}_1\frac{C_o}{2}\exp\left[-i(\Delta\beta_{01} - \gamma)z\right], \tag{23}$$

$$i\frac{d\hat{A}_1(z)}{dz} = -\hat{A}_0\frac{C_o}{2}\exp\left[i(\Delta\beta_{01} - \gamma)z\right]. \tag{24}$$

Next, we make the following relabeling $\Delta\beta_{01} \equiv \Delta\beta$ and define the new absorption operators $\hat{A}_0 = \hat{b}_0\exp(-i\Delta\beta z/2)$ and $\hat{A}_1 = \hat{b}_1\exp(i\Delta\beta z/2)$. We rewrite the Heisenberg equations in terms of the $\hat{b}(z)$ operators,

$$i\frac{d\hat{b}_0}{dz} = -\frac{\Delta\beta}{2}\hat{b}_0 - \frac{C_o}{2}\hat{b}_1\exp(i\gamma z), \tag{25}$$

$$i\frac{d\hat{b}_1}{dz} = \frac{\Delta\beta}{2}\hat{b}_1 - \frac{C_o}{2}\hat{b}_0\exp(-i\gamma z). \tag{26}$$

It is interesting to note that $\hat{a}_{0,1}$ and $\hat{b}_{0,1}$ are, after the spatial RWA, the same operators, except a global phase, that is $\hat{a}_{0,1} = \hat{b}_{0,1} \exp(i\bar{\beta}z)$, $\bar{\beta} = (\beta_0 + \beta_1)/2$. Finally, as mentioned above, we can write the above equation in a matrix form acting on the single-photon state $|L_b(z)\rangle = b_0(z)|1_0\rangle + b_1(z)|1_1\rangle$, where $|L(z)\rangle = |L_b(z)\rangle \exp(i\bar{\beta}z)$, that is,

$$i\hbar \frac{\partial}{\partial z}|L_b(z)\rangle = -\frac{\hbar}{2} \begin{pmatrix} \Delta\beta & C_o \exp(i\gamma z) \\ C_o \exp(-i\gamma z) & -\Delta\beta \end{pmatrix} |L_b(z)\rangle =$$

$$= -\frac{\hbar}{2}B \begin{pmatrix} \cos\alpha & \sin\alpha \exp(i\gamma z) \\ \sin\alpha \exp(-i\gamma z) & -\cos\alpha \end{pmatrix} |L_b(z)\rangle = -\hat{M}(z)|L_b(z)\rangle, \tag{27}$$

with $\cos\alpha = \Delta\beta/B$, $\sin\alpha = C_o/B$, and $B = \sqrt{\Delta\beta^2 + C_o^2}$. This equation is just the spatial equivalent of Equation (2). A Hamiltonian operator is replaced by a momentum operator. Obviously, the dynamic properties are identical under the formal changes: $\omega \leftrightarrow \gamma$, $\gamma_p B_o \leftrightarrow B$, $\theta \leftrightarrow \alpha$. Note that a rotating equivalent vector $(\gamma_p \mathbf{B})_{eq} \equiv \mathbf{B} = B(\sin\alpha \cos\gamma z, -\sin\alpha \sin\gamma z, \cos\alpha)$ is obtained. In short, we have achieved a photonic simulator of aspin-magnetic field coupling. However, we must stress some limitations for this simulator. The momentum operator $\hat{M}$ defined by (27) and simulating the coupling spin-magnetic field only is valid under the spatial RWA, that is the values of $\Delta\beta$ and γ have to be high and not too different. Therefore, we will be able to simulate the interaction with magnetic fields with a large z-component and oscillating with a frequency ω of the same order as the term $\gamma_p B_o \cos\theta$.

Finally, it is interesting to obtain a constant momentum operator by applying a unitary transformation (rotating reference system) to the quantum state $|L_b(z)\rangle$, that is,

$$|l(z)\rangle = \hat{U}(z,\gamma)|L_b(z)\rangle = \exp(-i\gamma\sigma_z z/2)|L_b(z)\rangle. \tag{28}$$

Therefore, by using $|L_b(z)\rangle = \exp(i\gamma\sigma_z z/2)|l(z)\rangle$ in Equation (27), we obtain, after a certain, but straightforward calculation, the following momentum operator for the state $|l(z)\rangle$:

$$-\hat{M}'|l(z)\rangle = \frac{\hbar}{2} \begin{pmatrix} -\Delta\beta + \gamma & -C_o \\ -C_o & \Delta\beta - \gamma \end{pmatrix} |l(z)\rangle = -\frac{\hbar}{2}D_o \begin{pmatrix} \cos\alpha' & \sin\alpha' \\ \sin\alpha' & -\cos\alpha' \end{pmatrix} |l(z)\rangle, \tag{29}$$

where $\cos\alpha' = (\Delta\beta - \gamma)/D_o$, $\sin\alpha' = C_o/D_o$, and $D_o = \sqrt{(\Delta\beta - \gamma)^2 + C_o^2}$. As shown later, these results are important for simulating both dynamical and topological properties. By comparison between Equations (5) and (29), we obtain the following simulation parameters:

$$(\omega - \gamma_p B_o \cos\theta) \leftrightarrow (\Delta\beta - \gamma), \quad \gamma_p B_o \sin\theta \leftrightarrow C_o. \tag{30}$$

Next, it is interesting to compute the eigenstates $|l(z)\rangle$ and eigenvalues β_l of the momentum operator $\hat{M}'$ given by Equation (6), that is $|l(z)\rangle = \exp\left(\frac{i}{\hbar}p_l z\right)|l(0)\rangle = \exp(i\beta_l z)|l(0)\rangle$. After a standard calculation, we have:

$$|l_+(0)\rangle = \cos\frac{\alpha'}{2}|1_0\rangle + \sin\frac{\alpha'}{2}|1_1\rangle, \quad |l_-(0)\rangle = \sin\frac{\alpha'}{2}|1_0\rangle - \cos\frac{\alpha'}{2}|1_1\rangle, \tag{31}$$

with eigenvalues $\beta_\pm = \pm D_o/2$, that is linear momentums $p_\pm = \pm\frac{\hbar}{2}D_o = \pm p_o$. Therefore, by taking into account Equation (28), the full evolved states $|L_b(z)\rangle$ can be easily obtained. Note that we are simulating the quantum state $\Psi(t)$ given by Equation (4), and thus, for example, quantum processing based on NMR could be simulated by this photonic device. For the sake of expositional convenience, we will return to this question in the next subsection.

3.4. Photonic Simulation of Light-Matter Interaction

For the case of light-matter interaction, we have a more direct photonic simulation by using a two-mode planar guide perturbed by an integrated optical grating. Thus, by defining $\Delta\beta = \beta_0 - \beta_1$ and $\bar{\beta} = (\beta_0 + \beta_1)/2$ and for the sake of simulation purposes, choosing an initial phase $\alpha_0 = \pi$, the momentum operator defined in Equation (20) can be rewritten as follows:

$$\hat{M}|L(z)\rangle = \hbar\left[(\bar{\beta} + C_+)\mathbf{I} + (\frac{\Delta\beta}{2} - C_-)\sigma_z - C_o\cos(\gamma z)\sigma_x\right]|L(z)\rangle, \tag{32}$$

where $C_\pm = (C_{00}(z) \pm C_{11}(z))/2$. As in the mechanical case, these terms could be zero if the optical modes $e_{n,m}(x)$ and perturbation $\Delta\epsilon(x)$ have a suitable parity, that is even or odd modes along with an odd perturbation. Next, we perform the following relevant change in the vector representation of the single-photon state $|L(z)\rangle \equiv (l_0(z)\exp(i\gamma z/2), l_1(z)\exp(-i\gamma z/2))^t \equiv \exp(i\gamma\sigma_z z/2)|l(z)\rangle$, with t indicating the transpose. By inserting this state into the above equation, using $\hat{M} = -i\hbar\partial/\partial z$, and neglecting terms that are rapidly oscillating, that is $\exp(\pm iq\gamma z)$ with $q = 1/2, 1, 3/2$, the following momentum operator is obtained for the state $|l(z)\rangle \equiv (l_0(z), l_1(z))^t$:

$$\hat{M}'|l(z)\rangle \approx \left(\hbar\bar{\beta}\mathbf{I} + \frac{\hbar\delta_s}{2}\sigma_z - \frac{\hbar C_o}{2}\sigma_x\right)|l(z)\rangle, \tag{33}$$

with $\delta_s = \Delta\beta - \gamma$. This momentum operator simulates the Hamiltonian given by Equation (13), where the following simulation parameters are obtained:

$$\bar{\omega} \leftrightarrow -\bar{\beta}, \quad \delta = (\omega_0 - \omega) \leftrightarrow \delta_s = (\Delta\beta - \gamma), \quad \Omega \leftrightarrow C_o, \tag{34}$$

with $\delta \leftrightarrow \delta_s$ the detuning simulation parameter and $\Omega \leftrightarrow C_o$ the Rabi frequency simulation parameter. Hence, this photonic device simulates light-matter interaction under a spatial RWA. Obviously, all temporal dynamics obtained by light-matter interaction can be simulated by means of this photonic device, as for example Rabi oscillations, logic gates for one-qubit transformations, and so on. In order to make clear these possibilities, let us consider the synchronous (or resonant) case, that is $\Delta\beta = \gamma$. The momentum operator is thus simplified, and the solutions can be easily obtained. In matrix form, the solution of the equation above is given by:

$$\begin{pmatrix} l_0(z) \\ l_1(z) \end{pmatrix} = \begin{pmatrix} \cos\frac{C_o}{2}z & -i\sin\frac{C_o}{2}z \\ -i\sin\frac{C_o}{2}z & \cos\frac{C_o}{2}z \end{pmatrix} \begin{pmatrix} l_0(0) \\ l_1(0) \end{pmatrix} \equiv X(\Theta/2)\begin{pmatrix} l_0(0) \\ l_1(0) \end{pmatrix}, \tag{35}$$

with $\Theta = C_o z$ and where an irrelevant global phase $e^{i\bar{\beta}z}$ has been omitted. As a simple example, if we choose a length of the grating (interaction length) $z = 2\pi/C_o$, then a X quantum logic gate is implemented. We must stress that these are transformations corresponding to the so-called Θ-pulses in atom-light temporal interaction for computing purposes [22]. For example, given an input state $|1_0\rangle$, the state propagating along the optical grating will be:

$$|L(z)\rangle = l_0(z)|1_0\rangle + l_1(z)|1_1\rangle. \tag{36}$$

On the other hand, for the case $|\delta_s| \ll C_o$ and by omitting again the global phase $e^{i\bar{\beta}z}$, we obtain, as a solution to Equation (33), a phase gate, that is,

$$\begin{pmatrix} l_0(z) \\ l_1(z) \end{pmatrix} = \begin{pmatrix} \exp(i\frac{\delta_s}{2}z) & 0 \\ 0 & \exp(-i\frac{\delta_s}{2}z) \end{pmatrix} \begin{pmatrix} l_0(0) \\ l_1(0) \end{pmatrix} \equiv Z(\Phi/2), \tag{37}$$

where $\Phi = \delta_s z$. Therefore, a Z quantum logic gate is obtained from Equation (37) if $\delta_s z/2 = \pi/2$, an S-gate if $\delta_s z/2 = \pi/4$, a T-gate if $\delta_s z/2 = \pi/8$, and so on. Moreover, by using an optical grating implementing a transformation $X(\Theta/2)$ and two phase gates $Z(\Phi/2)$, a ZXZ-factorization of SU(2) is obtained; therefore, any unitary transformation can be implemented with the photonic device, and consequently, any one-qubit can be generated.

Next, we present solutions for the asynchronous case, that is $\Delta\beta \neq \gamma$, which provides a most general solution and will be very useful to obtain geometric phases. By using standard methods to solve a linear equation system, the general solution of Equation (33) is given by:

$$\begin{pmatrix} l_0(z) \\ l_1(z) \end{pmatrix} = \begin{pmatrix} \cos\frac{\delta_r}{2}z + i\frac{\delta_s}{\delta_r}\sin\frac{\delta_r}{2}z & -i\frac{C_o}{\delta_r}\sin\frac{\delta_r}{2}z \\ -i\frac{C_o}{\delta_r}\sin\frac{\delta_r}{2}z & \cos\frac{\delta_r}{2}z - i\frac{\delta_s}{\delta_r}\sin\frac{\delta_r}{2}z \end{pmatrix} \begin{pmatrix} l_0(0) \\ l_1(0) \end{pmatrix}, \tag{38}$$

with $\delta_r = \sqrt{\delta_s^2 + C_o^2}$ and where, once more, the irrelevant global phase $e^{i\bar{\beta}z}$ is omitted. Note that for $\delta_s = 0$, the matrix given by Equation (35) is recovered. Likewise, under the condition $|\delta_s| \ll C_o$, the matrix solution given by Equation (37) is obtained. We must recall that all these transformations are obtained in a spatial rotating reference system defined by Equation (28) and induced by the RWA approximation, as occurs in atom-light temporal interactions. Likewise, we must stress that by using the simulation parameters given by Equation (34), mechanical solutions for light-matter (atom) interaction are directly obtained. Finally, it is also easy to check that the same solutions are obtained for the spin-magnetic field interaction simulation given by Equation (29); therefore, such an interaction has the same quantum processing properties as the atom-optical field interaction.

Finally, it is worth paying attention to the problem of properly initializing a quantum state. For that, let us consider an SPDC (Spontaneous Parametric Down Conversion) source of biphotons $|1_{k_a}1_{k_b}\rangle$, that is twin photons excited in two spatial modes propagating along directions k_a and k_b. Photon $|1_{k_b}\rangle$ is directed towards an APD device, and the other one $|1_{k_a}\rangle$ is directed to the prism-waveguide coupler, as shown in Figure 1a). The direction k_a is chosen in such a way that, for example, if the fundamental mode of the planar guide is excited, that is $k_a = k_0$, then we obtain the single-photon state (or register) $|1_0\rangle$. Alternatively, direction $k_a = k_1$ can be chosen to excite the mode e_1 of the planar guide, and thus, the single-photon state (or register) $|1_1\rangle$ is obtained. If we had channel waveguides, a similar procedure can be used, for example before the channel waveguide, there would be a planar waveguide with a prism-waveguide coupler and, next, an integrated lens or a similar integrated optical element [15] in such a way that the excited mode is focused onto the channel waveguide to excite the desired single-photon state. Next, by using optical gratings, different quantum transformations can be performed, and consequently, a general output state $|L\rangle = a_0(z)|1_{k_0}\rangle + a_1(z)|1_{k_1}\rangle$ is obtained. Finally, at a distance z, another prism-waveguide coupler can be placed at the end of the device to detect the quantum state. Photon detections can be made by using coincidences between the output photon and the photon $|1_{k_b}\rangle$ reaching the APD.

3.5. Implementation of Photonic Simulators

In this subsection, we present more complex simulators by using several integrated optical gratings along with other integrated components as Directional Couplers (DCs) made with Single-Mode (SMWs) or Two-Mode channel Waveguides (TMWs). We present a simulator of the interaction between an atom with four levels and Θ-pulses, next the interaction of a particle (or physical system) with spin $3/2$ and a magnetic field, and finally, an arbitrary unitary transformation SU(4) implemented with optical gratings what allows reducing the number of paths by half or even to a quarter, which can be generalized to SU(N) transformations, which can be of interest for photonic simulators.

Let us consider, as the first example, the Optical Grating (OG) studied above, but with a number $d = 4$ of collinear guided modes whose propagation constants are β_j with $j = 1, 2, 3, 4$. This simple device can simulate a four-level atom, which can be used to implement a CNOT logic gate operation under atom-laser interaction [22]. The optical grating fulfills $\gamma = \beta_2 - \beta_3$; therefore, it will produce an efficient transition between Modes 2 and 3 if $z = \pi/C_o$ (π-pulse in atom-laser interaction) according to

Equation (35). We can identify each single-photon state excited in each optical mode as a computational state (two-qubit single-photon [23]); thus, we have the state $|L\rangle = c_{00}|00\rangle + c_{01}|01\rangle + c_{10}|10\rangle + c_{11}|11\rangle$. When this state goes through the optical grating, the output state is $|L\rangle = c_{00}|00\rangle + c_{01}|01\rangle + c_{11}|10\rangle + c_{10}|11\rangle$, that is a CNOT operation is obtained. Obviously, this is not the best way to implement logic gates for two-qubits, but it makes clear how OGs can be used to simulate interaction between an atom with several levels and the so-called Θ-pulses.

The second example consists of using single-mode channel guides coupled by OGs in order to simulate the interaction between a particle (or physical system) with spin s and a magnetic field. This can be implemented by using a number $N=2s+1$ of optical modes excited in N channel waveguides and coupled by OGs, as shown in Figure 2 for the particular case of four guides (spin $s=3/2$). We must stress that coupling between parallel SMWs can be also described by the Heisenberg quantum equations given by Equation (18). In order to simulate spin-magnetic field interaction, either the strength of the OGs or the separation between consecutive SMWs has to be adjusted, and therefore, proper coupling strength values $C_{j,j+1}$ with $j=1,...N-1$, between consecutive guides, are obtained. Likewise, the SMWs have to be designed with different propagation constants β_j (asynchronous guides) by changing, for example, the depth of each channel waveguide. Finally, it is well known in integrated optics that for asynchronous DCs assisted by optical gratings, the coupling due to the overlapping fields is negligible. As a particular example, let us consider spin $s = 3/2$. The general equation system is given by:

$$-i\hbar \frac{d\hat{a}_j(z)}{dz} = \hbar\tilde{\beta}_j(z)\hat{a}_j(z) + \hbar \sum_{j'\neq j} C_{jj'}(z)\hat{a}_{j'}(z). \tag{39}$$

with $(|j - j'| < 2)$, that is coupling only exists between consecutive channel guides. Next, by using the same procedure followed for the case of spin-1/2, we obtain, after a long, but straightforward calculation, the following equation system:

$$-i\hbar\frac{\partial}{\partial z}|L_b(z)\rangle = -\Delta_o\mathbb{I}_{4x4}+$$

$$+\begin{pmatrix} 3\Delta_o & C_{12}\exp(i\gamma z) & 0 & 0 \\ C_{21}\exp(-i\gamma z) & \Delta_o & C_{23}\exp(i\gamma z) & 0 \\ 0 & C_{32}\exp(-i\gamma z) & -\Delta_o & C_{34}\exp(i\gamma z) \\ 0 & 0 & C_{43}\exp(-i\gamma z) & 3\Delta_o \end{pmatrix}|L_b(z)\rangle \tag{40}$$

where $\mathbb{I}_{4x4}$ is the identity matrix and the following propagation constants are used: β_1, $\beta_2=\beta_1-\Delta_o$, $\beta_3=\beta_1-2\Delta_o$, $\beta_4=\beta_1-3\Delta_o$. These values can be achieved by adjusting, for example, the width of the channel waveguides. On the other hand, the coupling coefficients for the gratings fulfil the relationships $C_{12}=C_{21}=C_{34}=C_{43}=\sqrt{3}C_o$ and $C_{23}=C_{32}=2C_o$, which can be achieved by adjusting the separation between consecutive guides with optical modes $e_j(x,y)$, $j=1,2,3,4$. In our case, Guides $2-3$ are closer than $1-2$ and $3-4$ because the coupling between Guides $2 and 3$ is larger, as shown in Figure 2. Finally, by defining $\cos\theta = \Delta_o/\Gamma$ and $\sin\theta = C_o/\Gamma$, with $\Gamma = (\Delta_o^2 + C_o^2)^{1/2}$, and a fictitious magnetic field $\mathbf{B}_f = (\sin\theta\cos\gamma z, -\sin\theta\sin\gamma z, \cos\theta)$, we can write the above equation system as follows:

$$-i\hbar\frac{\partial}{\partial z}|L_b(z)\rangle = \left\{-\hbar\Delta_o\mathbb{I}_{4x4} + \frac{\hbar}{2}\Gamma\,\mathbf{J}(3/2)\,\mathbf{B}_f\right\}|L_b(z)\rangle \tag{41}$$

where $\mathbf{J}(3/2) = (J_x(3/2), J_y(3/2), J_z(3/2))$ are the spin matrices for spin $s = 3/2$. In short, we constructed a photonic simulator for interaction between a spin-3/2 particle, or physical system, and a periodic magnetic field.

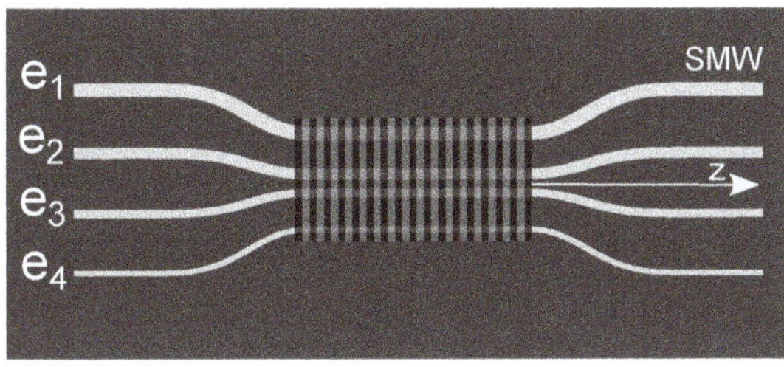

Figure 2. Four non-synchronous Single-Mode channel Waveguides (SMWs) assisted by an optical grating to simulate the interaction between a spin $s = 3/2$ particle and an oscillating magnetic field.

Next, we present a general unitary transformation SU(4) by using optical gratings. With this example, we want to show that OGs allow reducing the number of paths required for constructing a simulator. Indeed, an SU(4) simulator is formed by four paths implemented by SMWs with optical modes $e_{00}^{(j)}(x, y)$ (j=1, ...,4), six DCs, and six Phase Shifters (PSs), as shown in Figure 3 (bottom). However, if OGs are used, we only need two paths, that is TMWs with modes $e_{00}^{(j)}(x, y)$, $e_{10}^{(j)}(x, y)$ ($j = 1, 2$) and six OGs, as shown in Figure 3 (up). In this case, we also use Selective Directional Couplers (SDCs) (a SDC always performs an X transformation in one mode, and the other mode undergoes an identity transformation). Note that the number of paths was reduced by half because OGs act on collinear modes, unlike directional couplers, which act on codirectional modes. Obviously, if we had used OGs with four modes, we could reduce the number of paths by 1/4. Overall, we will be able to reduce the number of paths by $1/d$ if we use OGs with a number d of collinear modes. Ultimately, we can take advantage of the OGs to increase integration and thus to implement more flexible and scalable photonic simulators such as for example boson sampling ones [24], where the number of paths would be reduced by half. In short, the number N of paths of any required SU(N) transformation [25] could be reduced up to N/d.

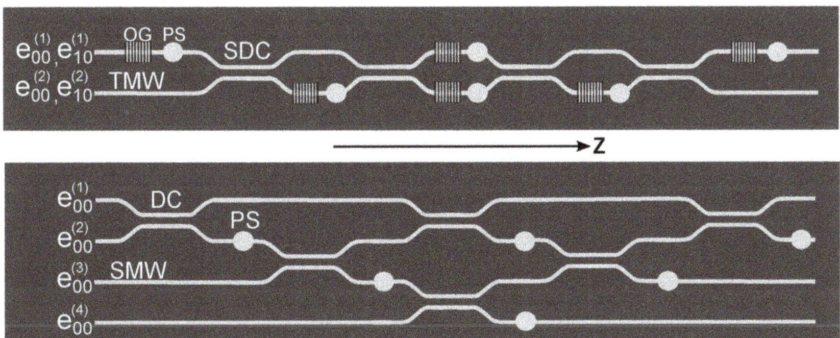

Figure 3. Standard implementation (bottom) of an arbitrary unitary transformation SU(4) by using Single-Mode channel Waveguides (SMWs) and Directional Couplers (DCs) (**bottom**). Alternative implementation (**top**) by using Two-Mode channel Waveguides (TMWs), Selective Directional Couplers (SDCs) and Optical Gratings (OGs). PS, Phase Shifter.

Finally, it is important to indicate that quantum photonic devices have their own limitations [4]. Thus, the difficulty to implement two-qubit logic gates is well known, which is, at present, an important drawback in general purpose quantum photonic computation. However, quantum photonic simulation, or simply quantum photonic computation, for specific purposes can provide efficient technological solutions, particularly if new degrees of freedom are incorporated. In our case, we have just shown that integrated OGs allow processing with several collinear modes, which improves the optical integration for high-dimensional problems, that is it provides a moderate increase of the on-chip flexibility and scalability for photon based quantum simulation. Moreover, OGs enable simulating quantum devices under variable perturbations and, in particular, periodic perturbations, which usually appear in quantum systems interacting with fields like spin ($N=2s+1$ modes)-magnetic field interaction, atom (d modes)-optical field interaction, and so on.

4. Quantum Geometric Phases

So far, we have simulated the dynamics of the spin-magnetic and light-matter interaction systems with a photonic device. However, these quantum devices also generate topological or geometric phases besides the dynamic phases. Geometric phases are precisely due to geometric properties as was originally proven by Berry in his seminal work about an adiabatic quantum system [26]. Later, these geometric phases were generalized by Aharonov and Anandan [27] to non-adiabatic processes, and their calculation is made by using the geometric properties of the projective Hilbert space. Finally, a useful extension to geometric phases associated with non-cyclic circuits on the projective Hilbert space was also proposed [28]. On the other hand, topological phases in optics have also been extensively studied in bulk devices with polarization modes [29] and also in integrated optics with spatial modes [30]. At present, geometrical phases have regained interest for their possible application to geometric quantum computation [31,32]. Non-adiabatic spatial propagation on the Hilbert space generate the geometric phase known as the Aharonov–Anandan (AA) phase. It is well known that the spin-magnetic field interaction, as for example NMR, produces AA phases. We must stress that the geometric phase for a two-dimensional projective Hilbert space can be calculated in a geometric way as $\phi_g = (1/2)\Omega(\mathcal{C})$, where $\Omega(\mathcal{C})$ is the solid angle subtended by the circuit $\mathcal{C}$ followed by the quantum state on the Bloch sphere. In this section, we prove that quantum geometrical phases can be obtained by an integrated photonic grating, and therefore, it simulates the geometric phases produced by both a spin-magnetic field system and an atom-optical field system.

4.1. Geometric Phases in Spin-Magnetic Field Photonic Simulation

Let us consider the eigenstates $|l_\pm(0)\rangle$ given by Equation (31). Spatial propagation of these states is given, according to Equations (28) and (31), by $|L_b(z)\rangle \equiv |L_\pm(z)\rangle = \exp(i\gamma\sigma_z z/2)|l_\pm(z)\rangle = \exp(\frac{ip_\pm}{\hbar}z)\exp(i\gamma\sigma_z z/2)|l_\pm(0)\rangle$. For instance, let us take $|L_+(z)\rangle$ after a propagation distance $z = \nu\Lambda$, that is for a photonic integrated grating with a length $\nu\Lambda$, where $\Lambda = 2\pi/\gamma$ and ν is the number of cycles taken, then we have:

$$|L_+(\nu\Lambda)\rangle = \exp\left(\frac{ip_+}{\hbar}\nu\Lambda\right)\left(\cos\frac{\alpha'}{2}\exp(i\nu\pi)|1_0\rangle + \sin\frac{\alpha'}{2}\exp(-i\nu\pi)|1_1\rangle\right) \tag{42}$$

By following the same procedure for the state $|L_-(z = \nu\Lambda)\rangle$ and taking into account that ν is an integer, we obtain:

$$|L_\pm(\nu\Lambda)\rangle = \exp\left(\frac{ip_\pm}{\hbar}\nu\Lambda \pm i\nu\pi\right)|l_\pm(0)\rangle = \exp(i\phi^\pm))|l_\pm(0)\rangle. \tag{43}$$

The phases $\phi^\pm$ obtained above are the full phases acquired by the states $|L_\pm\rangle$ after ν cycles in the optical grating, that is,

$$\phi^\pm = \nu(\pm p_o\Lambda/\hbar \pm \pi) \tag{44}$$

A photonic Bloch sphere is shown on the left in Figure 4 where each point corresponds to a single-photon state given by Equation (19). For comparison purposes, an NMR Bloch sphere is also shown on the right. A single-photon state propagating along z can be represented by the following general expression $|L(z)\rangle = c_0(z)|1_0\rangle + c_1(z)|1_1\rangle$; therefore, each point (x,y,z) of the photonic Bloch sphere is defined as follows: $x = 2\operatorname{Re}c_0(z)c_1(z)$, $y = 2\operatorname{Im}c_0(z)c_1(z)$, $z = |c_0(z)|^2 - |c_1(z)|^2$, where Re and Im stand for real and imaginary parts, respectively.

It is easy to check that eigenstates $|L_\pm(z)\rangle$ follow the curves C_c and C_c' corresponding to spherical caps, as shown in Figure 4. The solid angle subtended by a spherical cap with an angular extension α' is given by $\Omega(C) = 2\pi(1 - \cos\alpha')$.

As the full phase ϕ can be decomposed into a dynamical part ϕ_d and a geometric one ϕ_g, then the geometric phase can be obtained from the relationship:

$$\phi_\mathrm{g}^\pm = \phi_\pm - \phi_\mathrm{d}^\pm \tag{45}$$

We first compute the dynamical phase, that is $\phi_\mathrm{d}^\pm = \frac{1}{\hbar}\int_0^{v\Lambda}\langle L_\pm(z)|\hat{M}(z)|L_\pm(z)\rangle dz$; therefore, by taking into account the transformation (28) and expression (29) for $\hat{M}'$, we obtain:

$$\begin{aligned}
\phi_\mathrm{d}^\pm &= \tfrac{1}{\hbar}\int_0^{n\Lambda}\langle L_\pm(z)|\hat{M}(z)|L_\pm(z)\rangle dz = \tfrac{1}{\hbar}\int_0^{v\Lambda}\langle l_\pm(0)|\hat{M}(0)|l_\pm(0)\rangle dz = \\
&= \tfrac{1}{\hbar}\int_0^{v\Lambda}\left(\langle l_\pm(0)|\hat{M}'|l_\pm(0)\rangle - \tfrac{\gamma}{2}\langle l_\pm(0)|\sigma_z|l_\pm(0)\rangle\right)dz = \\
&= \tfrac{p_\pm}{\hbar}v\Lambda \pm v\pi\cos\alpha',
\end{aligned} \tag{46}$$

and by taking into account the total phase given by Equation (43), the geometrical phases acquired by the eigenstates are given by:

$$\phi_\mathrm{g}^\pm = \pm v\pi(1 - \cos\alpha') = \pm v\pi\left(1 - \frac{\Delta\beta - \gamma}{\sqrt{(\Delta\beta - \gamma)^2 + C_0^2}}\right). \tag{47}$$

Note that the geometric phase is half the solid angle subtended by the circuit, where the sign $\pm$ depends on the direction of rotation followed by the state on the Bloch sphere. The corresponding spherical cap circuits C_c and C_c' are shown in Figure 4. Moreover, the geometric phase depends on the device parameters, that is $\Delta\beta, \gamma, C_0$. Likewise, it is important to indicate that the dynamical phase is of order ω^1 (note that $\beta \propto \omega$), but the geometrical phase is of order ω^0; therefore, the geometric phase is less sensitive to errors in the distance propagation.

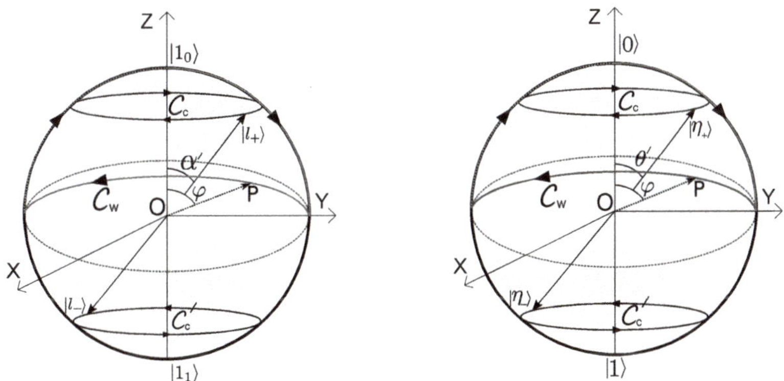

Figure 4. On the left, a photonic Bloch sphere shows the evolution of the single-photon states $|l_+\rangle$ and $|l_-\rangle$ (spherical cap circuits C_c and C_c'). Likewise, a spherical wedge circuit C_w is shown. On the right, the simulated mechanical Bloch sphere for an interacting atom-field system is shown, with similar spherical cup circuits for states $|\eta_+\rangle$ and $|\eta_-\rangle$ and also a spherical wedge circuit C_w.

In short, a single-photon state acquires an AA geometric phase under propagation in an integrated photonic grating. The same expression is found for a spin-1/2 particle in a magnetic field; therefore, topological simulations can be made. Thus, by applying the simulation parameters given by Equation (30), geometric phases can be obtained.

4.2. Elimination of the Dynamical Phase in Spin-Magnetic Field Photonic Simulation

It is well known in the mechanical case that the geometric phase is hidden in the spin-magnetic field interaction because it is combined with the dynamical phase within $\phi^{\pm}$; therefore, the dynamical phase has to be eliminated in order to take advantage of the properties of a geometric phase. We present a photonic solution, which is similar to the one used in the mechanical case, that is if the quantum state, after evolution under a first Hamiltonian $\hat{H}_1 = \hat{H}$ for time T, finds a second Hamiltonian $\hat{H}_2 = -\hat{H}$, then the dynamical phase are mutually canceled; however, the eigenstates do not change, therefore neither does the geometrical phase.

In the photonic case, we have to find a new momentum operator, that is a new integrated optical grating, such as $\hat{M}_2 = -\hat{M}$. We assume that such a new optical grating has the same frequency γ; therefore, we have the same rotating system, that is the same transformation (28). Accordingly, the condition for eliminating the dynamical phase is obtained from the operator $\hat{M}'$, that is,

$$\hat{M}'_2 = \frac{\hbar}{2} \begin{pmatrix} \Delta\beta' - \gamma & C'_o \\ C'_o & -\Delta\beta' + \gamma \end{pmatrix} = -\hat{M}' = -\frac{\hbar}{2} \begin{pmatrix} \Delta\beta - \gamma & C_o \\ C_o & -\Delta\beta + \gamma \end{pmatrix}; \tag{48}$$

therefore, $C'_o = -C_o$ and $(\Delta\beta - \gamma) = -(\Delta\beta' - \gamma)$. The first condition can be achieved by introducing an initial phase in the second grating, that is $\Delta\epsilon(x, z) = \Delta\epsilon(x)\cos(\gamma z + \pi)$, and the second one is achieved if $\Delta\beta' = 2\gamma - \Delta\beta$. These results indicate that we need an additional grating with a new difference between propagation constants (linear momentum of the photon) $\Delta\beta' = (\beta'_0 - \beta'_1)$. In Figure 5 the system for eliminating the dynamical phase is shown . Therefore, the total phases are $\phi_{\pm} = \nu(\mp p\Lambda/\hbar - \pi)$, and the dynamical phases are:

$$\phi_{\mathrm{d}}^{\pm} = \frac{p_{\pm}}{\hbar}\nu\Lambda \mp \nu\pi\cos\alpha' \tag{49}$$

Therefore, the total dynamical phase is $\Phi_d = 0$, and the total geometrical phase after the single-photon state propagates through the two integrated gratings is twice the value acquired in the first grating, that is,

$$\Phi_g^{\pm} = \mp 2\nu\pi(1 - \cos\alpha') = \mp\Phi_g. \tag{50}$$

Alternatively, propagation constants can be unchanged, and the grating frequency can be modified, that is $\gamma' = 2\Delta\beta - \gamma$; however, in this case, the transformation (28) must be applied with the factor γ'. In short, we eliminated the dynamical phase; therefore, these results could be used for implementing logic gates or transformations based on topological phases, which are much more insensitive to fabrication errors, unlike dynamical phases, which as mentioned are of order ω. With these results, robust P-gates (Phase gates) can be designed; thus, the following transformation is implemented between the eigenstates:

$$P = \begin{pmatrix} \exp(-i2\nu\pi\cos\alpha') & 0 \\ 0 & \exp(i2\nu\pi\cos\alpha') \end{pmatrix} = \exp(-i\Phi_g\sigma_z) \tag{51}$$

Note that for $4\nu\cos\alpha' = \pm 1$, a Z-gate is obtained, and for $4\nu\cos\alpha' = \pm 1/2$, an S-gate is implemented, and so on.

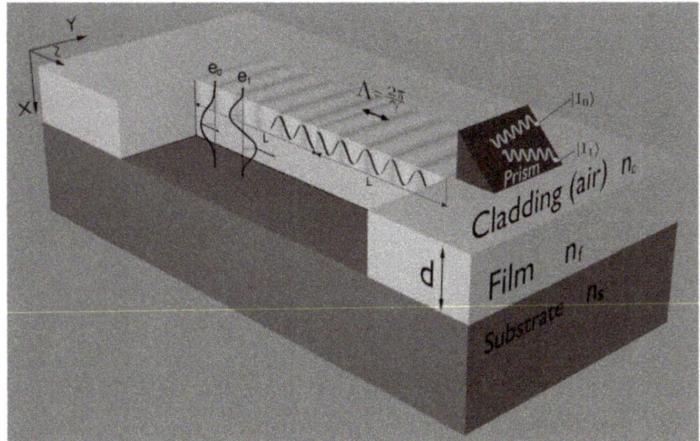

Figure 5. Elimination of the dynamical phase by using two consecutive integrated optical gratings. The second grating has an initial phase π. Likewise, a prism is used to make a projective measure of states $|1_0\rangle$ and $|1_1\rangle$ for obtaining the probabilities P_0 and P_1 and, therefore, the geometric phase Φ_g.

4.3. Geometric Phases with Other Quantum States

On the other hand, the eigenstates given by Equation (31) can be written as single-photon states excited in rotated optical modes, that is $e_+(x,y) = \cos\frac{\alpha'}{2}e_0(x,y) + \sin\frac{\alpha'}{2}e_1(x,y)$ and $e_-(x,y) = -\sin\frac{\alpha'}{2}e_0(x,y) + \cos\frac{\alpha'}{2}e_1(x,y)$; therefore:

$$|1_+(0)\rangle = \cos\frac{\alpha'}{2}|1_0\rangle + \sin\frac{\alpha'}{2}|1_1\rangle = |1_+\rangle, \quad |1_-(0)\rangle = -\sin\frac{\alpha'}{2}|1_0\rangle + \cos\frac{\alpha'}{2}|1_1\rangle = |1_-\rangle \tag{52}$$

The elimination of the dynamical phase means that these eigenstates have undergone the transformation $e^{\pm i\Phi_g}|1_\pm\rangle$; therefore, the following formal relationships can be written:

$$e^{i\Phi_g}|1_+\rangle = e^{i\Phi_g}\hat{a}_+^\dagger|0_\pm\rangle, \quad e^{-i\Phi_g}|1_-\rangle = e^{-i\Phi_g}\hat{a}_-^\dagger|0_\pm\rangle \tag{53}$$

with $\Phi_g = -2\nu\pi(1-\cos\alpha')$. Accordingly, the following transformations, induced by geometric phases, for the absorption operators are obtained:

$$\hat{a}_\pm(z=n\Lambda) = e^{\pm i\Phi_g}\,\hat{a}_\pm(0) \tag{54}$$

This result can be used for obtaining geometric phases of other quantum light states. As an example, we present two states, that is the number photon state or Fock state $|n_+\rangle$ and the coherent state $|\alpha_+\rangle$, where subindex $+$ indicates that the state is excited in the optical mode $e_+(x,y)$. The Fock state under propagation becomes:

$$|n_+(0)\rangle = \frac{1}{\sqrt{n_+!}}(\hat{a}_+(0)^\dagger)^{n_+}|0\rangle \longrightarrow \frac{1}{\sqrt{n_+!}}e^{in_+\Phi_g}(\hat{a}_+^\dagger)^{n_+}|0\rangle = e^{in_+\Phi_g}|n_+\rangle \tag{55}$$

Therefore, the quantum state has acquired a geometric phase $n_+\Phi_g$. Likewise, the coherent state can be rewritten by using the complex displacement operator, that is,

$$|\alpha_+(0)\rangle = e^{(\alpha+\hat{a}_+^\dagger(0)-c.h.)}|0\rangle \longrightarrow e^{(\alpha+e^{i\Phi_g}\hat{a}_+^\dagger-c.h.)}|0\rangle = |e^{i\Phi_g}\alpha_+\rangle \tag{56}$$

Therefore, the geometric phase is Φ_g, that is the same as the one acquired by a single photon. The same procedure can be applied to any other quantum light state excited in the integrated photonic grating.

4.4. Optical Measurement of Geometric Phases

Finally, we present how to measure the geometric phase starting from the measurements of the single photon detection probability, which can be extracted by a prism-waveguide coupler [10], as shown in Figure 5. We focus on the spin-magnetic field interaction simulation case. By assuming that the input state is a single photon excited in the mode $e_0(x, y)$ and taking into account the relationships given by Equation (52), the following final state is obtained after the two gratings:

$$|L(z)\rangle = (\cos^2 \frac{\alpha'}{2} + \sin^2 \frac{\alpha'}{2} e^{i\Phi_g})|1_0\rangle + \cos \frac{\alpha'}{2} \sin \frac{\alpha'}{2}(1 - e^{i\Phi_g})|1_1\rangle \tag{57}$$

The probability of the detection of a photon in Mode 0, that is P_0, or in Mode 1, P_1, is a function of the geometric phase, that is,

$$P_1 = \frac{\sin^2 \alpha}{2}(1 - \cos \Phi_g), \quad P_0 = 1 - P_1 \tag{58}$$

If $\Phi_g = 2\pi$, then $P_1 = 0$ and $P_0 = 1$, and if $\Phi_g = \pi$, then $P_1 = \sin^2 \alpha$ and $P_0 = \cos^2 \alpha$. Therefore, from the measurement of P_1 and P_0, the phase Φ_g is obtained. In Figure 5 it is shown the projective measure of states $|1_0\rangle$ and $|1_1\rangle$ by a prism-waveguide coupler, which projects these state in different spatial directions. Finally, note that by using a coherent state, these probabilities are proportional to the intensity of the light, what can be called a semiclassical optical characterization, or in the most technical way, the geometric phase is also acquired by the classical fields, but would rigorously correspond to the so-called Hannay phase [33].

4.5. Geometric Phases in Light-Matter Photonic Simulation

Finally, we check that light-matter simulation can be also used to obtain geometric phases. For the sake of simplicity, we show geometric phases for wedge circuits, although more general cases can be studied. Let us consider an initial state $|L(0)\rangle = |1_0\rangle$, that is the point $(0, 0, 1)$ on the photonic Bloch sphere. Next, let us consider an asynchronous optical grating with $\delta_s \ll C_o$; therefore, according to Equation (37) (phase gate), for $\delta_s z_o = 3\pi/2$, we obtain $|L(z_o)\rangle = (1/\sqrt{2})(|1_0\rangle + i|1_1\rangle)$, that is the state reaches the Bloch sphere point $(0, 1, 0)$. Next, we consider that the grating has a greater coupling coefficient C_o, then the single-photon state is given by the expression:

$$|L(z)\rangle = \frac{1}{\sqrt{2}}[(\cos \frac{\delta_r z}{2} + b \sin \frac{\delta_r z}{2} + ia \sin \frac{\delta_r z}{2})|1_0\rangle + (a \sin \frac{\delta_r z}{2} + i \cos \frac{\delta_r z}{2} - ib \sin \frac{\delta_r z}{2})|1_1\rangle] \tag{59}$$

where $a = \delta_s/\delta_r$ and $b = C_o/\delta_r$. If we choose a distance $z = z_1$ such as $\delta_r z_1/2 = \pi/2$, then, after a certain calculation, the following state is obtained:

$$|L(z_1)\rangle = \frac{e^{i\phi_o}}{\sqrt{2}}(|1_0\rangle + i|1_1\rangle) \tag{60}$$

where $\phi_o = \text{atn}(a/b) = \text{atn}(\delta_s/C_o)$, that is $a = \sin \phi_o$ and $b = \cos \phi_o$. The state has reached the point $(0, -1, 0)$ of the photonic Bloch sphere. Now, we show that ϕ_o is a geometric phase. Indeed, for the sake of symmetry, the state before reaching the above state has crossed the meridian $y = 0$ when $\delta_r z/2 = \pi/4$, that is, the state:

$$|L'\rangle = \frac{(1 + b + ia)}{2}|1_0\rangle + \frac{(a + i(1 - b))}{2}|1_1\rangle \equiv m_0 e^{i\epsilon_0}|1_0\rangle + m_1 e^{i\epsilon_1}|1_1\rangle. \tag{61}$$

It is easy to prove that both states have the same phase, that is $\epsilon_0 = \epsilon_1$, and the modulus is given by $m_0 = \sin(\phi_o/2)$ and $m_1 = \cos(\phi_o/2)$; therefore, the state crosses the point $P = (\sin \phi_o, 0, \cos \phi_o)$ of the photonic Bloch sphere as indicated in Figure 4 for $\varphi = \phi_o$. Now, we must recall that partial cycles also generate geometric phases, which can be calculated by closing the end points of the open cycle by a geodesic line [28,30]. In our case, the corresponding geodesic lies along the meridian at the plane $x = 0$, from point $(0, -1, 0)$ to initial point $(0, 0, 1)$. In short, the state has followed a wedge circuit C_w, as shown in Figure 4. Now, we calculate the subtended solid angle by this wedge circuit. It is easy to check that a wedge circuit with angle ϕ_o subtends a solid angle $\Omega(C_w) = 2\phi_o$; therefore, the geometric phase is $\Phi_g = (1/2)\Omega(C_w) = \phi_o = \text{atn}(\delta_s/C_o)$, which is just the global phase obtained in Equation (60). It is worth underlining that in this case, the geometric phase is not hidden, and therefore, the dynamical phase does not have to be eliminated. Moreover, this geometric phase can be also obtained in a atom-optical field system out of resonance by using Θ-pulses, with the first optical field with low amplitude E_o (Z gate) and the next with a higher amplitude E_o. By using the simulation parameters given by Equation (34), the geometric phase would be $\Phi_g = \text{atn}(\delta/\Omega)$.

5. Conclusions

We propose a quantum photonic device based on integrated optical gratings in a two-mode slab guide to simulate the interaction between external fields and atoms. By using single-photon states, we study the simulations of a spin-(1/2)-magnetic field system, as for example nuclear magnetic resonance, and a two-level atom-optical field system corresponding to light-matter simulation. Both dynamical and geometric properties are simulated, in particular the geometric phases obtained by the mentioned systems. We prove that dynamical properties can be simulated for a wide range of cases with practical interest, although in the spin-(1/2)-magnetic field system it is restricted to relatively high values of frequency and magnetic field amplitude B_0. Overall, atom-optical field interaction does not present these restrictions. This study of integrated optical gratings opens up possibilities to more general simulations if several modes are used. Thus, spin (s)-magnetic field interaction simulations could be implemented by using a number $N = 2s+1$ of codirectional optical modes assisted by optical gratings; multilevel atom (n)-optical field interaction can be simulated by using $N = n$ collinear optical modes coupled by optical gratings, which can in turn simulate, for example, two-qubit single photon logic gates, which has a high interest in quantum information systems; likewise, optical gratings allow interaction between d collinear modes, and thus, simulators based on N codirectional modes can reduce the number of paths used up to N/d, which improves the optical integration of the photonic simulator. On the other hand, AA geometric phases have been also obtained for both systems. The spin-(1/2)-magnetic field system requires dynamic phase cancellation, which is simulated by using two optical gratings; however, in the atom-optical field system, such cancellation is not required. Obviously, we must emphasize that although the proposed integrated photonic device is intended for quantum simulation, it can also be used to implement quantum operations and/or effects with the photonic device itself, such as logic gates, geometric phases, and so on, by using single-photon states or more general quantum states, as shown.

Author Contributions: All authors contributed equally to this work. All authors read and agreed to the published version of the manuscript.

Funding: Xunta de Galicia, Consellería de Educación, Universidades e FP, Grant GRC Number ED431C2018/11; Ministerio de Economía, Industria y Competitividad, Gobierno de España, Grant Number AYA2016-78773-C2-2-P.

Acknowledgments: One of the authors (G.M.C.) wishes to acknowledge the financial support by Xunta de Galicia, Consellería de Educación, Universidades e FP, by a predoctoral grant co-financed with the European Social Fund.

Conflicts of Interest: The authors declare no conflict of interest. The funders had no role in the design of the study; in the collection, analyses, or interpretation of data; in the writing of the manuscript; nor in the decision to publish the results.

Abbreviations

The following abbreviations are used in this manuscript:

RWA   Rotating Wave Approximation
NMR   Nuclear Magnetic Resonance
AA   Aharonov–Anandan
SPDC   Spontaneous Parametric Down Conversion
DC   Directional Coupler
SDC   Selective Directional Coupler
APD   Avalanche Photodiode
SMW   Single-Mode Waveguide
TMW   Two-Mode Waveguide
OG   Optical Grating

References

1. Huttner, B.; Serulnik, S.; Ben-Aryeh, Y. Quantum analysis of light propagation in a parametric amplifier. *Phys. Rev. A* **1990**, *42*, 5594–5600. [CrossRef]
2. Liñares, J.; Nistal, M. Quantization of coupled modes propagation in integrated optical waveguides. *J. Mod. Opt.* **2003**, *50*, 781–790. [CrossRef]
3. Johnson, T.H.; Clark, S.R.; Jarsch, D. What is a quantum simulator? *EPJ Quantum Technol.* **2014**, *1*, 10. [CrossRef]
4. Georgescu, I.; Ashhab, S.; Nori, F. Quantum simulation. *Rev. Mod. Phys.* **2014**, *86*, 153–185. [CrossRef]
5. Schaetz, T.; Monroe, C.R.; Esslinger, T. Focus on quantum simulation. *New J. Phys.* **2013**, *15*, 085009. [CrossRef]
6. Agarwal, G. *Quantum Optics*; Cambridge Unversity Press: Cambridge, UK, 2013.
7. Longui, S.; Della Valle, G. Anyons in one-dimensional lattices: A photonic realization. *Opt. Lett.* **2012**, *37*, 2160–2162. [CrossRef]
8. Le Bellac, M. *A Short Introduction to Quantum Information and Quantum Computation*; Cambridge University Press: Cambridge, UK, 2006.
9. Loudon, R. *The Quantum Theory of Light*; Oxford University Press: Oxford, UK, 2003.
10. Lee, D. *Electromagnetic Principles of Integrated Optics*; Wiley: New York, NY, USA, 1986.
11. Doerr, C. Silicon photonic integration in telecommunications. *Front. Phys.* **2015**, *3*, 7–22. [CrossRef]
12. Politi, A.; Matthews, J.C.F.; Thompson, M.G.; O'Brien, J.L. Integrated Quantum Photonics. *IEEE J. Sel. Top. Quantum Electron.* **2009**, *15*, 1673–1684. [CrossRef]
13. Wang, J.; Sciarrinom, F.; Laing, A.; Thompson, M. Integrated photonic quantum technologies. *Nat. Photonics* **2020**, *14*, 273–284. [CrossRef]
14. Campany, J.; Pérez, D. *Programmable Integrated Photoncs*; Oxford Unversity Press: Oxford, UK, 2020.
15. Liñares, J.; Montero, C.; Moreno, V.; Nistal, M.C.; Prieto, X.; Salgueiro, J.R.; Sotelo, D. Glass processing by ion exchange to fabricate integrated optical planar components: Applications. *Proc. SPIE* **2000**, *3936*, 227–238.
16. Righini, G.; Chiappini, A. Glass optical waveguides: A review of fabrication techniques. *Opt. Eng.* **2014**, *53*, 071819. [CrossRef]
17. Cacciari, I.; Brenci, M.; Falciai, R.; Conti, G.; Pelli, S.; Righini, G. Reproducibility of splicer-based long-period fiber gratings for gain equalization. *Front. Phys.* **2007**, *3*, 203–206. [CrossRef]
18. Yao, Y.; Liu, H.; Xue, L.; Liu, B.; Zhang, H. Design of lithium niobate phase-shifted Bragg grating for electro-optically tunable ultra-narrowbandwidth filtering. *Appl. Opt.* **2019**, *58*, 6770–6774. [CrossRef] [PubMed]
19. Ben-Aryeh, Y.; Serulnik, S. The quantum treatment of propagation in non-linear optical media by the use of temporal modes. *Phys. Lett. A* **1991**, *155*, 473–479. [CrossRef]
20. Liñares, J.; Nistal, M.; Barral, D. Quantization of Coupled 1d Vector Modes In Integrated Photonics Waveguides. *New J. Phys.* **2008**, *10*, 063023. [CrossRef]
21. Liñares, J.; Nistal, M.; Barral, D.; Moreno, V. Optical field-strength polarization of two-mode single-photon states. *Eur. J. Phys.* **2010**, *31*, 991–1005. [CrossRef]
22. Fox, M. *Quantum Optics*; Oxford Unversity Press: Oxford, UK, 2006.

23. Englert, B.; Kurtsiefer, C.; Weinfurter, H. Universal unitary gate for single-photon two-qubit states. *Phys. Rev. A* **2001**, *63*, 032302. [CrossRef]

24. Spring, J.; Metcalf, B.; Humphreys, P.; Kolthammer, W.; Jin, X.; Barbieri, M.; Datta, A.; Thomas-Peter, N.; Langford, N.K.; Kundys, D.; et al. Boson sampling on photonic chip. *Science* **2013**, *239*, 798–801. [CrossRef]

25. de Guise, H.; Di Matteo, O.; Sánchez-Soto, L. Simple factorization of unitary transformations. *Phys. Rev. A* **2018**, *97*, 022328. [CrossRef]

26. Berry, R. Quantal phase factors accompanying adiabatic change. *Proc. R. Soc. Lond. A* **1987**, *392*, 45–64.

27. Aharonov, Y.; Anandan, J. Phase change during a cyclic quantum evolution. *Phys. Rev. Lett.* **1987**, *58*, 1593–1596. [CrossRef]

28. Jordan, T. Berry phases for partial cycles. *Phys. Rep.* **1988**, *38*, 1590–1592. [CrossRef]

29. Bhandari, R. Polarization of light and topological phases. *Phys. Rep.* **1997**, *281*, 1–64. [CrossRef]

30. Liñares, J.; Nistal, M. Geometric phases in multidirectional electromagnetic coupling theory. *Phys. Lett. A* **1992**, *162*, 7–14. [CrossRef]

31. Ekert, A.; Ericsson, M.; Hayden, P.; Inamori, H.; Jones, J.; Oi, D.; Vedral, V. Geometric quantum computation. *J. Mod. Opt.* **2000**, *47*, 2501–2513. [CrossRef]

32. Sjöqvist, E. Geometric phases in quantum information. *Int. J. Quantum Chem.* **2015**, *115*, 1311–1326. [CrossRef]

33. Agarwal, T.; Simon, R. Berry phase, interference of light beams and the Hannay angle. *Phys. Rev. A* **1990**, *42*, 6924–6927. [CrossRef]

Publisher's Note: MDPI stays neutral with regard to jurisdictional claims in published maps and institutional affiliations.

Review

Towards Quantum 3D Imaging Devices

Cristoforo Abbattista [1], Leonardo Amoruso [1], Samuel Burri [2], Edoardo Charbon [2], Francesco Di Lena [3], Augusto Garuccio [3,4], Davide Giannella [3,4], Zdeněk Hradil [5], Michele Iacobellis [1], Gianlorenzo Massaro [3,4], Paul Mos [2], Libor Motka [5], Martin Paúr [5], Francesco V. Pepe [3,4,*], Michal Peterek [5], Isabella Petrelli [1], Jaroslav Řeháček [5], Francesca Santoro [1], Francesco Scattarella [3,4], Arin Ulku [2], Sergii Vasiukov [3], Michael Wayne [2], Claudio Bruschini [2,†], Milena D'Angelo [3,4,*,†], Maria Ieronymaki [1,†] and Bohumil Stoklasa [5,†]

[1] Planetek Hellas E.P.E., 44 Kifisias Avenue, 15125 Marousi, Greece; abbattista@planetek.gr (C.A.); amoruso@planetek.gr (L.A.); iacobellis@planetek.gr (M.I.); petrelli@planetek.gr (I.P.); santoro@planetek.gr (F.S.); ieronymaki@planetek.gr (M.I.)

[2] Ecole Polytechnique Fédérale de Lausanne (EPFL), 2002 Neuchâtel, Switzerland; samuel.burri@epfl.ch (S.B.); edoardo.charbon@epfl.ch (E.C.); paul.mos@epfl.ch (P.M.); arin.ulku@epfl.ch (A.U.); michael.wayne@epfl.ch (M.W.); claudio.bruschini@epfl.ch (C.B.)

[3] INFN, Sezione di Bari, 70125 Bari, Italy; francesco.dilena@ba.infn.it (F.D.L.); augusto.garuccio@uniba.it (A.G.); davide.giannella@uniba.it (D.G.); gianlorenzo.massaro@uniba.it (G.M.); francesco.scattarella@uniba.it (F.S.); sergii.vasiukov@ba.infn.it (S.V.)

[4] Dipartimento Interateneo di Fisica, Università degli Studi di Bari, 70126 Bari, Italy

[5] Department of Optics, Palacký University, 17. Listopadu 12, 77146 Olomouc, Czech Republic; hradil@optics.upol.cz (Z.H.); libor.motka@upol.cz (L.M.); martin.paur@upol.cz (M.P.); michal.peterek@upol.cz (M.P.); rehacek@optics.upol.cz (J.Ř.); bohumil.stoklasa@upol.cz (B.S.)

* Correspondence: francesco.pepe@ba.infn.it (F.V.P.); milena.dangelo@ba.infn.it (M.D.)

† Equal last author contribution.

Citation: Abbattista, C.; Amoruso, L.; Burri, S.; Charbon, E.; Di Lena, F.; Garuccio, A.; Giannella, D.; Hradil, Z.; Iacobellis, M.; Massaro, G.; et al. Towards Quantum 3D Imaging Devices. *Appl. Sci.* **2021**, *11*, 6414. https://doi.org/10.3390/app11146414

Academic Editors: Maria Bondani, Alessia Allevi and Stefano Olivares

Received: 5 May 2021
Accepted: 7 July 2021
Published: 12 July 2021

Publisher's Note: MDPI stays neutral with regard to jurisdictional claims in published maps and institutional affiliations.

Abstract: We review the advancement of the research toward the design and implementation of quantum plenoptic cameras, radically novel 3D imaging devices that exploit both momentum–position entanglement and photon–number correlations to provide the typical refocusing and ultra-fast, scanning-free, 3D imaging capability of plenoptic devices, along with dramatically enhanced performances, unattainable in standard plenoptic cameras: diffraction-limited resolution, large depth of focus, and ultra-low noise. To further increase the volumetric resolution beyond the Rayleigh diffraction limit, and achieve the quantum limit, we are also developing dedicated protocols based on quantum Fisher information. However, for the quantum advantages of the proposed devices to be effective and appealing to end-users, two main challenges need to be tackled. First, due to the large number of frames required for correlation measurements to provide an acceptable signal-to-noise ratio, quantum plenoptic imaging (QPI) would require, if implemented with commercially available high-resolution cameras, acquisition times ranging from tens of seconds to a few minutes. Second, the elaboration of this large amount of data, in order to retrieve 3D images or refocusing 2D images, requires high-performance and time-consuming computation. To address these challenges, we are developing high-resolution single-photon avalanche photodiode (SPAD) arrays and high-performance low-level programming of ultra-fast electronics, combined with compressive sensing and quantum tomography algorithms, with the aim to reduce both the acquisition and the elaboration time by two orders of magnitude. Routes toward exploitation of the QPI devices will also be discussed.

Keywords: quantum imaging; plenoptic imaging; quantum correlations; SPAD arrays; quantum fisher information; compressive sensing

1. Introduction

Fast, high-resolution, and low-noise 3D imaging is highly required in the most diverse fields, from space imaging to biomedical microscopy, security, industrial inspection, and cultural heritage [1–5]. In this context, conventional plenoptic imaging represents one of the most promising techniques in the field of 3D imaging, due to its superb temporal resolution:

Appl. Sci. **2021**, *11*, 6414. https://doi.org/10.3390/app11146414   https://www.mdpi.com/journal/applsci

3D imaging is realized in a single shot, for seven frames per second at 30 M pixel resolution, and 180 frames per second for 1M pixel resolution [5]; no multiple sensors, near-field techniques, time-consuming scanning or interferometric techniques are required. However, conventional plenoptic imaging entails a loss of resolution which is often unacceptable. Our strategy to break such limitation consists of combining a radically new and foundational approach with last-generation hardware and software solutions. The fundamental idea is to exploit the information stored in correlations of light by using novel sensors and measurement protocols to implement a very ambitious task: high speed (10–100 fps) quantum plenoptic imaging (QPI) characterized by ultra-low noise and an unprecedented combination of resolution and depth-of-field. The developed imaging technique aims at becoming the first practically usable and properly "quantum" imaging technique that surpasses the intrinsic limits of classical imaging modalities. In addition to the foundational interest, the quantum character of the technique allows for extracting information on 3D images from correlations of light at very low photon fluxes, thus reducing the scene exposure to illumination. The interest in QPI is specially motivated by the potential advantages with respect to other established 3D imaging techniques. Actually, other methods require, unlike QPI, either delicate interferometric measurements, as in digital holographic microscopy [6,7], or phase retrieval algorithms, as in Fourier ptychography [8], or fast pulsed illumination, as in time-of-flight (TOF) imaging [9–14]. Moreover, QPI provides the basis of a scanning-free microscopy modality, overcoming the drawbacks of the confocal methods [15].

In view of the deployment of quantum plenoptic cameras suitable for real-world applications, the crucial challenge is represented by the reduction of both the acquisition and data elaboration times. In fact, a typical complication arises in quantum imaging modalities based on correlation measurements: the reconstruction of the correlation function encoding the desired image requires collecting a large number of frames (30,000–50,000 in the first experimental demonstration of the refocusing capability of correlation plenoptic imaging [16]), which must be individually read and stored before elaborating the output. Therefore, to get an estimate of the total time required to form a quantum plenoptic image, the data reading and transmission times must be added to the acquisition time of the employed sensor. This problem is addressed by an interdisciplinary approach, involving the development of ultrafast single-photon sensor systems, based on SPAD arrays [17–22], the optimization of circuit electronics to collect and manage the high number of frames (e.g., by GPU) [23,24], the development of dedicated algorithms (compressive sensing, machine learning, quantum tomography) to achieve the desired SNR with a minimal number of acquisitions [25–28]. Finally, the performances of QPI will be further enhanced by a novel approach to imaging based on quantum Fisher information [29,30]. Treating the physical model of plenoptic imaging in the view of quantum information theory brings new possibilities of improving the setup towards super-resolution capability in the object 3D space. Having the optimal set of optical setup parameters enables object reconstruction close to ultimate limits set by nature. In this article, we will combine a review of the state of the art in the aforementioned fields with the discussion on how they contribute to the development of QPI and its application.

The paper is organized as follows: in Section 2, we discuss the working principle and recent advances of Correlation Plenoptic Imaging (CPI), a technique that represents the direct forerunner of QPI; in Section 3, we present the hardware innovations currently investigated to reduce the acquisition times in CPI; in Section 4, we review the algorithmic solutions to improve QPI; in Section 5, we outline the perspectives of our future work in the context of Qu3D project; in Section 6, we discuss the relevance of our research. The work presented in this paper involves experts from three scientific research institutions, Istituto Nazionale di Fisica Nucleare (INFN, Italy), Palacky University Olomouc/Department of Optics (UPOL, Czechia), and Ecole polytechnique fédérale de Lausanne/Advanced Quantum Architecture Lab (EPFL, Switzerland), and from the industrial partner Planetek Hellas E.P.E. (PKH, Greece). The activity is carried within the project "Quantum 3D

Imaging at high speed and high resolution" (Qu3D), founded by the 2019 QuantERA call [31].

2. Plenoptic Imaging with Correlations: From Working Principle to Recent Advances

Quantum plenoptic cameras promise to offer the advantages of plenoptic imaging, primarily ultrafast and scanning-free 3D imaging and refocusing capability, with performances that are beyond reach for the classical counterpart. State-of-the-art plenoptic imaging devices are able to acquire multi-perspective images in a single shot [5]. Their working principle is based on the simultaneous measurement of both the spatial distribution and the propagation direction of light in a given scene. The acquired directional information translates into refocusing capability, augmentable depth of field (DOF), and parallel acquisition of multi-perspective 2D images, as required for fast 3D imaging.

In state-of-the-art plenoptic cameras [32], directional detection is achieved by inserting a microlens array between the main lens and the sensor of a standard digital camera (see Figure 1a). The sensor acquires composite information that allows identification of both the object point and the lens point where the detected light is coming from. However, the image resolution decreases with inverse proportionality to the gained directional information, for both structural (use of a microlens array) and fundamental (Gaussian limit) reasons; plenoptic imaging at the diffraction limit is thus considered to be unattainable in devices based on simple intensity measurement [5].

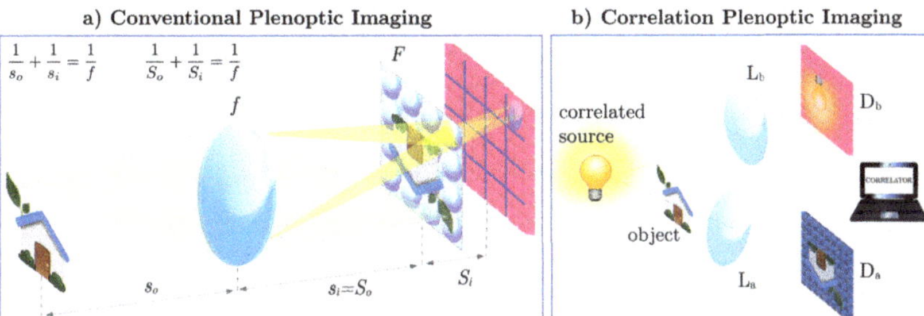

Figure 1. (**a**) the scheme of a conventional plenoptic imaging (PI) device: the image of the object is focused on a microlens array, while each microlens focuses an image of the main lens on the pixels behind. Such a configuration entails a loss of spatial resolution proportional to the gain in directional resolution; (**b**) shows the scheme of a correlation plenoptic imaging (CPI) setup, in which directional information is obtained by correlating the signals retrieved by a sensor on which the object is focused with a sensor that collects the image of the light source. The image in (**a**) is reproduced with the permission from Ref. [16], copyright American Physical Society, 2017.

Recently, the INFN group involved in Qu3D has proposed a novel technology, named Correlation Plenoptic Imaging (CPI), that enables overcoming the resolution drawback of current plenoptic devices, while keeping their advantages in terms of refocusing capability and 3D reconstruction [1,16,33]. CPI is based on either intensity correlation measurement or photon coincidence detection, according to the light source: actually, CPI can be based on the spatio-temporal correlations characterizing both chaotic sources [16,33] and entangled photon beams [34] to encode the spatial and directional information on two disjoint sensors, as shown in Figure 1b. CPI with chaotic light is based on the measurement of the correlation function

$$\Gamma(\boldsymbol{\rho}_a, \boldsymbol{\rho}_b) = \langle I_a(\boldsymbol{\rho}_a) I_b(\boldsymbol{\rho}_b) \rangle - \langle I_a(\boldsymbol{\rho}_a) \rangle \langle I_b(\boldsymbol{\rho}_b) \rangle, \tag{1}$$

where $\langle \dots \rangle$ denotes the average on the source statistics, and $I_j \rho_j$ ($j = a, b$) are the intensities propagated by the beam j and registered in correspondence of point ρ_j on the sensor D_j. Experimentally, the statistical averages are replaced by time averages, obtained by retrieving a collection of frames, simultaneously acquired by the two detectors. In CPI devices, the correlation function encodes combined information on the distribution of light on two reference planes, one of which corresponds to the "object plane" that would be focused on the sensor in a standard imaging setup, placed at a distance s_o from the focusing element. In general, given an object placed at a distance s from the focusing element, and characterized by the light intensity distribution $\mathcal{A}(\rho)$, its images are encoded in the function $\Gamma(\rho_a, \rho_b)$, in the geometrical-optics limit, as

$$\Gamma(\rho_a, \rho_b) \sim \mathcal{A}^n \left(\frac{s}{s_o} \frac{\rho_a}{M} + \left(1 - \frac{s}{s_o} \right) \frac{\rho_b}{M_L} \right), \tag{2}$$

where M and M_L are the magnifications of the images of the reference object plane and of the focusing element, respectively, while the power n is equal to 1 or 2, according to whether the object lies in only one [16,33,35] or both [36,37] optical paths.

Experimental CPI based on pseudo-thermal light is shown in Figure 2c, where both the acquired out-of-focus image and the corresponding refocused image are shown [16]. It was demonstrated in this proof of principle that CPI is characterized by diffraction-limited resolution on the object plane focused on the sensor. Details on the resolution limits are shown in Figure 2a, where one can observe an even more striking effect: thanks to its intrinsic coherent nature, CPI enables an unprecedented combination of resolution and DOF [16]. However, the low-noise sCMOS camera employed in the experiment, working at 50 fps at full resolution, requires several minutes to acquire 30,000 frames used to reconstruct the plenoptic correlation function, and a standard workstation has taken over 10 h for elaborating the acquired data and perform refocusing. The resulting image was also rather noisy, due to the well-known resolution vs. noise compromise of chaotic light ghost imaging that keeps also affecting CPI [35].

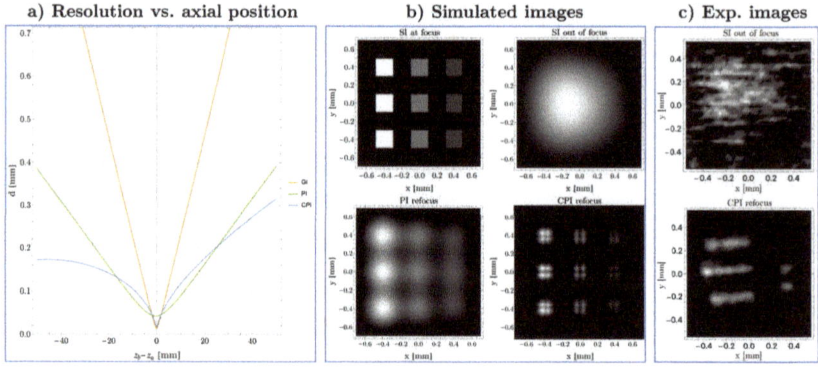

Figure 2. (**a**) shows the resolution limits, as a function of the longitudinal position, of the image of a double-slit mask with center-to-center distance d equal to twice the slit width; here, CPI outperforms both conventional imaging and standard PI with 3×3 directional resolution. The evident asymmetry of the CPI curve is due to the existence of two planes in which the object is focused: one at $z_b = z_a$ and one at $z_b = 0$ (see [16]). Plots in (**b**) show a result of a simulation: the target is moved from the focused plane (top left) to an out-of-focus plane (top right). Starting from this position, we show the results of PI refocusing with 3×3 directional resolution (bottom left) and the CPI refocusing (bottom right); (**c**) shows the results of an experiment [16] in which the standard image of a triple slit was completely blurred (top), while the image obtained by CPI (bottom) was made fully visible by exploiting information on light direction. Plots in (**c**) are reproduced with the permission from Ref. [16], copyright American Physical Society, 2017.

We are addressing these issues by employing two kinds of sources:

- Chaotic light sources, such as pseudothermal light, natural light, LEDs and gas lamps, and even fluorescent samples, operated either in the high-intensity regime or in the "two-photon" regime, in which an average of two photons per coherence area propagates in the setup. Chaotic light sources are well known to be characterized by EPR-like correlations in both momentum and position variables [38,39], to be exploited in an optimal way to retrieve an accurate plenoptic correlation function in the shortest possible time. In order to efficiently retrieve spatio-temporal correlations, tight filtering of the source can be necessary to match the filtered source coherence time with the response time of the SPAD arrays that can be as low as 1 ns. Alternatively, pseudorandom sources with a controllable coherence time, made by impinging laser light on a fast-switching digital micromirror device (DMD), can be employed. Interestingly, recent studies have shown that, in the case of chaotic light illumination, the plenoptic properties of the correlation function do not need to rely strictly on ghost imaging: correlations can be measured between any two planes where ordinary images (see Figure 1b) are formed [35]. This discovery has led to the intriguing result that the SNR of CPI improves when ghost imaging of the object is replaced by standard imaging [40]. In particular, excellent noise performances are expected in the case of images of birefringent objects placed between crossed polarizers. This kind of source is particularly relevant in view of applications in fields like biomedical imaging (cornea, zebrafish, invertebrates, biological phantoms such as starch dispersions), security (distance detection, DOF extension), and satellite imaging.

- Momentum–position entangled beams, generated by spontaneous parametric down-conversion (SPDC), which have the potential to combine QPI with sub-shot noise imaging [41], thus enabling high-SNR imaging of low-absorbing samples, a challenging issue in both biomedical imaging and security.

The design of both quantum plenoptic devices are currently undergoing optimization by implementing a novel protocol that enables further mitigating the resolution vs. DOF compromise with respect to the one shown in Figure 2a: this protocol is based on the observation that, for any given resolution, the DOF can be maximized by correlating the standard images of two arbitrary planes, chosen in the surrounding of the object of interest, instead of imaging the focusing element [36]. Moreover, we are investigating the possibility to merge quantum plenoptic imaging with the measurement protocols developed in the context of differential ghost imaging [42].

In the following section, we will discuss all the technical solutions, both on the hardware and on the software side that we are investigating to speed up the process of generating a correlation plenoptic image. The problematic aspects can be clarified by a description of the practical process of CPI imaging that is made of the following steps:

1. Collecting N_f synchronized pairs of frames from a sensor D_a, with resolution $N_{ax} \times N_{ay}$, and D_b, with resolution $N_{bx} \times N_{by}$. Synchronization of two separate digital sensors entails technical complications that can be overcome by using two disjoint parts of the same sensor [16]. The total acquisition time is

$$T = N_f(\tau_{exp} + \tau_d), \tag{3}$$

   where τ_{exp} is the frame exposure time that must be shorter than the coherence time of impinging light in order to exploit the maximal information on intensity fluctuations, while τ_d is the dead time between subsequent frames, usually fixed by the employed sensors.

2. Each acquired frame is transferred to a computer to be processed. This step can occur either progressively during the capture of the subsequent frames, or at the end of the acquisition process.

3. The collected frames are used to obtain an estimate of the correlation of intensity fluctuations (1) as

$$\Gamma(\boldsymbol{\rho}_a, \boldsymbol{\rho}_b) \simeq \frac{1}{N_f} \sum_{k=1}^{N_f} I_a^{(k)}(\boldsymbol{\rho}_a) I_b^{(k)}(\boldsymbol{\rho}_b) - \frac{1}{N_f} \sum_{k=1}^{N_f} I_a^{(k)}(\boldsymbol{\rho}_a) \frac{1}{N_f} \sum_{k'=1}^{N_f} I_b^{(k')}(\boldsymbol{\rho}_b), \tag{4}$$

where $I_j^{(k)}(\boldsymbol{\rho}_j)$, with $j = a, b$, is the intensity measured in frame k in correspondence of the pixel on the detector D_j centered on the coordinate $\boldsymbol{\rho}_j$. In this way, an information initially encoded in $N_f(N_{ux}N_{uy} + N_{bx}N_{by})$ numbers is used to reconstruct a correlation function determined by $N_{ax}N_{ay}N_{bx}N_{by}$ values.

The accuracy of the correlation function estimate (4) increases with the number of frames like $\sqrt{N_f}$ [40]. However, increasing N_f also linearly extends the total acquisition time T. Therefore, the combination of (1) fast and low-noise sensors, (2) methods to extract a good quality signal from smaller number of frames, and (3) tools for efficient data transfer and elaboration, is crucial in order to speed-up the acquisition process and make the CPI technology ready for real-world applications.

3. Hardware Speedup: Advanced Sensors and Ultra-Fast Computing Platforms

To improve the performances of CPI in terms of acquisition speed and data elaboration time, we are employing dedicated advanced sensors and ultra-fast computing platforms. In this section, we describe the details of the implementation and the perspectives on these fields.

3.1. SPAD Arrays as High-Resolution Time-Resolved Sensors

A relevant part of the speedup that we are seeking is determined by replacing commercial high-resolution sensors, like scientific CMOS and EMCCD cameras, with sensors based on cutting-edge technology such as single-photon avalanche diode (SPAD) arrays. A SPAD is basically a photodiode which is reversely biased above its breakdown voltage, so that a single photon which impinges onto its photosensitive area can create an electron–hole pair, triggering in turn an avalanche of secondary carriers and developing a large current on a very short timescale (picoseconds) [17,18]. This operation regime is known as Geiger mode. The SPAD output voltage is sensed by an electronic circuit and directly converted into a digital signal, further processed to store the binary information that a photon arrived, and/or the photon time of arrival. In essence, a SPAD can be seen as a photon-to-digital conversion device with exquisite temporal precision. SPADs can also be gated, in order to be sensitive only within temporal windows as short as a few nanoseconds, as shown in Figure 3. Individual SPADs can nowadays be used as the building blocks of large arrays, with each pixel circuit containing both the SPAD and the immediate photon processing logic and interconnect. Several CMOS processes are readily available and allow for tailoring both the key SPAD performance metrics and the overall sensor or imager architecture [43,44]. Sensitivity and fill-factor have for some time lagged behind those of their scientific CMOS or EMCCD counterparts but have been substantially catching up in recent years.

Based on the requirements of QPI, we have chosen to employ the SwisSPAD2 array developed by the AQUA laboratory group of EPFL, characterized by a 512×512 pixel resolution (see Figure 4), which is one of the widest and most advanced SPAD arrays to date [20,22]. The sensor is internally organized as two halves of 256×512 pixels to reduce load and skew on signal lines and enable faster operation. It is a purely binary gated imager, i.e., each pixel records either a 0 (no photon) or a 1 (one or more photons) for each frame, with basically zero readout noise. The sensor is controlled by an FPGA generating the control signals for the gating circuitry and readout sequence and collecting the pixel detection results. In the FPGA, the resulting one-bit images can be further processed, e.g., accumulated into multi-bit images, before being sent to a computer/GPU for analysis and storage. The maximum frame rate is 97.7 kfps, and the native fill factor of 10.5% can be improved by 4–5 times, for collimated light, by means of a microlens array [21] (higher values are expected from simulation after optimization); the photon detection probability is 50% (25%) at 520 nm (700 nm) and 6.5 V excess bias. The device is also characterized by low noise (typically less than 100 cps average Dark Count Rate per pixel at room temperature, with a median value about 10 times lower) and advanced circuitry for nanosecond gating.

A detailed comparison of SwissSPAD2 with other large-format CMOS SPAD imaging cameras is presented in Ref. [22].

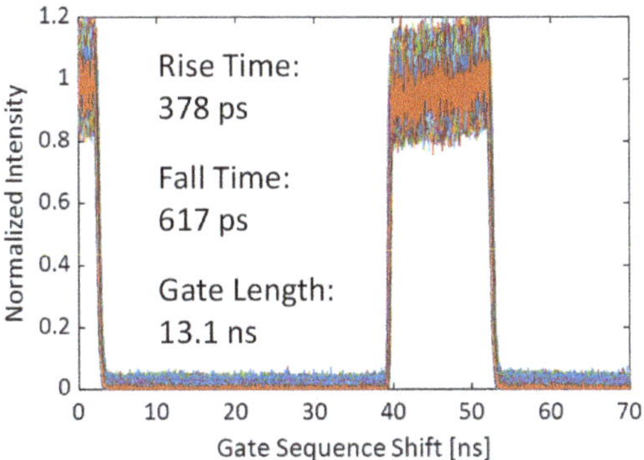

Figure 3. SwissSPAD2 gate window profile. The transition times and the gate width are annotated in the figure. The gate width is user-programmable, and the minimum gate width in the internal laser trigger mode is 10.8 ns. The image is reproduced with permission from Ref. [45], copyright The Authors, published by IOP Publishing Ltd., 2020.

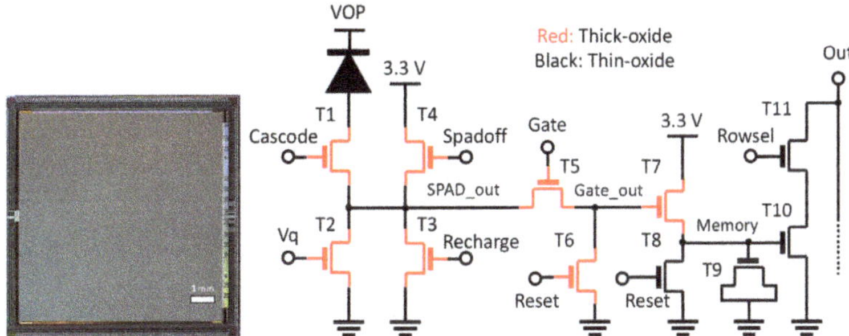

Figure 4. SwissSPAD2 photomicrograph (**left**) and pixel schematics (**right**). The pixel consists of 11 NMOS transistors, 7 with thick-oxide, and 4 with thin-oxide gate. The pixel stores a binary photon count in its memory capacitor. The in-pixel gate defines the time window, with respect to a 20 MHz external trigger signal, in which the pixel is sensitive to photons. The image is reproduced with permission from Ref. [22], copyright IEEE, 2018.

Currently, we are using the available version of SwissSPAD2, with a 512×256 pixel resolution, to generate sequences of frames and store them in an on-board 2 GB memory, before transferring them to a computer by a standard USB3 connection, which can be done using existing hardware. We are integrating this sensor in the prototype of chaotic-light base quantum plenoptic camera in a way that two disjoint halves of the sensor (of 256×256 pixels each) are used for retrieving the images of the two reference planes. The high speed of this sensor is expected to reduce the acquisition time of quantum plenoptic images by two order of magnitudes with respect to the first CPI experiment [16], in which the region of interest on the sensor was made of 700×700 pixels for the spatial measurement side, and 600×600 for the directional measurement side; 2×2 binning on both sides during acquisition and a subsequent 10×10 binning on the directional side led to the effective spatial resolution of 350 pixels, and angular resolution of 25×25 pixels.

We are also working towards several further optimizations of the sensor system, e.g., by developing gating for noise reduction in QPI devices. In order to employ the full sensor (512×512 pixels), we are implementing a synchronization mechanism for a pair of imagers by means of two FPGAs, so as to operate on a common time-base at the nanosecond level (to this end, two control circuits shall operate from a single clock and have a direct communication link). Finally, the SwissSPAD2 arrays are being integrated with a fast communication interface in order to speed up data transfer and make it possible to deliver, in a sustained way, full binary frame sequences to a GPU. The latter will run advanced algorithms for data pre-processing, image reconstruction and optimization, as we will discuss in more detail in the next subsection.

3.2. Computational Hardware Platform

In a QPI device integrated with SwissSPAD2, the acquired data rate for a single frame acquisition can be estimated to 26 Gb/s, which is beyond the reach of standard data buses. This poses great challenges in both hardware and software design. Our approach is to employ the expertise of PKH to seek careful design of electronic interconnections (buses) between sensor control electronics and processing device, theoretical refinement, and optimization of algorithms (e.g., compressive sensing [46,47]), porting to an efficient computational environment, and design of a specific acquisition electronics for optimizing data flow from the light sensors to a dedicated processing system, able to guarantee the required computational performance (e.g., exploiting GPUs or FPGA) [23,24].

The introduction of an embedded data acquisition- and processing board, integrating a GPU, aims at data pre-processing, thus significantly reducing the amount of data to be transferred to (and saved on) an external workstation. GPUs exploit a highly parallel elaboration paradigm, enabling to design algorithms that run in parallel on hundreds or thousands of cores and to make them available on embedded devices. A great advantage of GPUs is programmability: many standard tools exist (e.g., OpenCL and CUDA) that allow fast and efficient design of complex algorithms that can be injected on the fly in the GPU memory for accomplishing tasks ranging from simple filtering to advanced machine learning. Efforts are made to design a processing matrix, so that each line and column of the sensor will be managed by a dedicated portion of the heterogeneous processing platform (CPU/GPU/FPGA): the pixel series processor. Those dedicated units will be interconnected to one another to cooperate for implementing algorithms that require distributed processing on a very small scale.

The embedded acquisition-processing board is designed to best fit to SPAD array and SW application needs. The optimal system design will be evaluated by considering theoretical algorithms, engineered SW implementations, HW set-up, and HW/SW trade-off. We will identify a preliminary set of possible configurations and perform a trade-off by comparing overall performances, considering the requirements in data quality, processing speed, costs, complexity, etc.

Based on the challenging objectives to be achieved, a preliminary analysis based on COTS (Commercial-Off-The-Shelf) solutions was performed, in order to identify a set of accelerating devices addressing QPI requirements in terms of computing capability and portability. The option offered by the NVIDIA Jetson Xavier AGX board shown in Figure 5 is considered a promising candidate to achieve our goal. This device indeed offers an encouraging performance/integration ratio with low power usage and very interesting computing capabilities. Its main characteristics are reported in Table 1.

Considering the listed HW capabilities, despite the ARM processor having a limited computing power when compared to high-end desktop processors, it allows for leveraging multi-core capabilities for implementing that part of the code that will feed the quite powerful GPU device on board. In addition, the foreseen optimization strategy will require a dedicated implementation able to consider the maximum amount of memory of 16 GB available, shared with the GPU. In addition, the solution to be developed should take into account the bandwidth of about 136 GB/s of the on-board memory, which may represent a limiting factor when GPU and CPU exchange buffers. An implementation based on the CUDA framework—over OpenCL or other technologies—will be preferred to best use the NVIDIA device. Finally, given the capability of the NVDLA Engines to perform multiplications and accumulations in a very fast way, we consider it interesting to perform an assessment of how to exploit these devices for implementing the QPI-specific correlation functions and/or other multiplication/sum intensive computations.

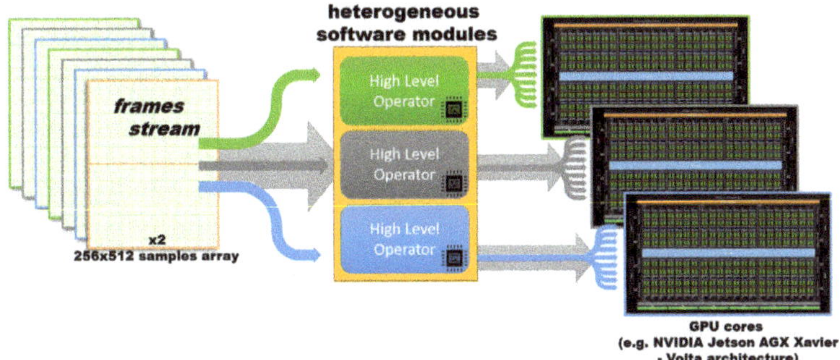

Figure 5. Qu3D scenario for GPU parallel processing based on NVIDIA Volta architecture as provided by an NVIDIA Jetson AGX Xavier device.

Table 1. NVIDIA Jetson Xavier AGX: Technical Specifications.

	Jetson Xavier AGX
GPU	512-core NVIDIA Volta™ GPU with 64 Tensor Cores
CPU	8-core NVIDIA Carmel Arm®v8.2 64-bit CPU 8MB L2 + 4MB L3
Memory	16 GB 256-bit LPDDR4x 136.5GB/s
PCIE	$1 \times 8 + 1 \times 4 + 1 \times 2 + 2 \times 1$ (PCIe Gen4, Root Port & Endpoint)
DL Accelerator	$2\times$ NVDLA Engines
Vision Accelerator	7-Way VLIW Vision Processor
Connectivity	10/100/1000 BASE-T Ethernet

Along with the assessment of a dedicated HW solution, further optimizations applicable at the algorithmic level have been considered in order to enrich the engineered device with a highly customized algorithmic workflow able to exploit the peculiarities of the CPI technique and its related input data. More specifically, we are analyzing those steps of QPI processing that appear as more computationally demanding, thus representing a bottleneck for performances. To this end, tailored reshaping operations applied over the original three-dimensional multi-frame structure of the input data were explored, to facilitate the development of a parallelized elaboration paradigm for evaluating the CPI-related correlation function. In addition, the peculiar feature of the input dataset to be acquired by SPAD sensors as one-bit images will be valued through dedicated implementations able to gain from intensive multiplication/sum math among binary-valued variables.

4. Quantum and Classical Image-Processing Algorithms

Further reduction of the acquisition speed and the optimization of the elaboration time is addressed by exploiting dedicated quantum and classical image processing, as well as novel mathematical methods coming on one hand from compressive sensing, and on the other hand from quantum tomography and quantum Fisher information.

4.1. Compressive Sensing

In order to reduce the amount of required data (at present 10^3–10^4 frames) by at least one order of magnitude, we are exploring different approaches. First, we are investigating the opportunity of implementing compression techniques for improving bus bandwidth utilization, thus acting on data transfer optimization. Data compression may also rely on manipulation of raw input data to determine only the relevant information in a sort of information bottleneck paradigm, in which software nodes in a lattice provide their contribution to a probable reconstruction of the actual decompressed data. This is very similar to artificial neural network structures, where the network stores a representation of a phenomenon and returns a response based on some similarity rating versus the observed data. Moreover, the availability of advanced processing technologies allows the investigation and implementation of alternative techniques able to exploit the sparsity of the retrieved signal to reconstruct information from a heavily sub-sampled signal, by compressive sensing techniques [46,47].

As in other correlation imaging techniques, such as ghost imaging (GI), in the CPI protocol, the object image is reconstructed by performing correlation measurements between intensities at two disjoint detectors. Katz et al. [48] demonstrated that conventional GI offers the possibility to perform compressive sensing (CS) boosting the recovered image quality.

CS theory asserts the possibility of recovering the object image from far fewer samples than required by the Nyquist–Shannon sampling theorem, and it relies on two main principles: sparsity of the signal (once expressed in a proper basis) and incoherence between the sensing matrix and the sparsity basis.

In conventional GI, the transmission measured for each speckle pattern represents a projection of the object image and CS finds the sparsest image among all the possible images consistent with the projections. In practice, CS reconstruction algorithms solve a convex optimization problem, looking for the image which minimizes the L_1-norm in the sparse basis among the ones compatible with the bucket measurements, see Refs. [49–52] for a review.

We are developing a novel protocol reducing the number of measurements required for image recovery by an order of magnitude. Once properly refocused, a single acquisition can be fed into the compressive sensing algorithm several times thus exploiting the plenoptic properties of the acquired data and increasing the signal-to-noise ratio of the final refocused image. We tested the CS-CPI algorithm with numerical simulations, as summarized in Figure 6. In (a), a double-slit mask is reconstructed by correlations measurements considering $N = 6000$ frames; in (b), the standard reconstruction is repeated considering only the 10% of available frames, while, in (c), the CS reconstruction using the same reduced set of measured data. In addition to a data-fidelity term corresponding to a linear regression, we penalized the L_1-norm of the reconstructed image to account for its sparsity in the $2D - DCT$ domain. The resulting optimization problem is known as the LASSO (Least Absolute Shrinkage and Selection Operator) [53]. We employed the coordinate descent algorithm to efficiently solve it, and we set the regularization parameter, controlling the degree of sparsity of the estimated coefficients, by cross-validation. In this proof-of-concept experiment, we simply use Pearson's correlation coefficient to measure the similarity between the reconstructions obtained using the restricted dataset and the image obtained considering all the $N = 6000$ frames.

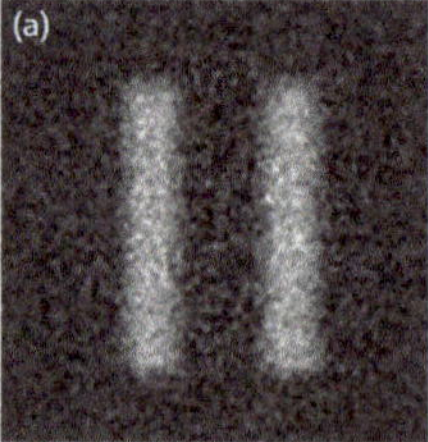
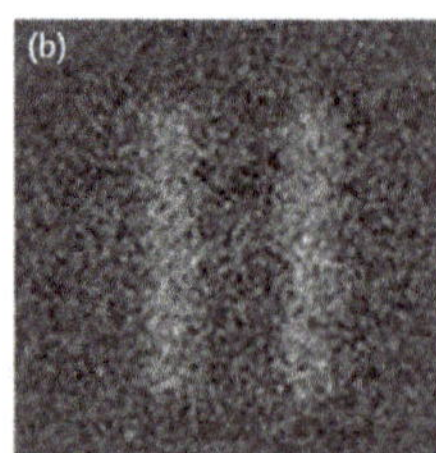
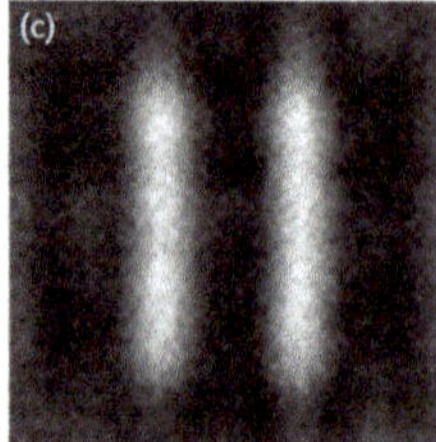

Figure 6. (**a**) double-slit image reconstruction obtained by correlations measurements, considering $N = 6000$ frames and two detectors characterized by a 128×128 and a 10×10 pixel resolution; (**b**) the standard reconstruction is repeated considering only the 10% of the available frames, chosen randomly; (**c**) compressive sensing reconstruction using the same dataset as in (**b**). While in the first case, the Pearson's correlation coefficient is $r_{red} = 0.55$; in the latter case, the coefficient is increased to $r_{CS} = 0.81$.

4.2. Plenoptic Tomography

Novel reconstruction algorithm with the advantage of real 3D image lack of artifacts is based on the idea of recasting plenoptic imaging as an absorption tomography. The classical refocusing algorithm is based on superimposing images of the 3D scene from different viewpoints, which necessarily results in the contribution of out-of-focus parts of the scene, the effect responsible for the existence of the blurred part of 3D image reconstruction. The tomography approach is based on a different principle and provides sufficient axial and transversal resolution without artifacts. For this purpose, the object space is divided into voxels and the goal is to reconstruct absorption coefficients of each voxel. The measured correlation coefficient of a pair of points from the correlated planes is transformed to be linearly proportional to the attenuation along the ray path connecting those two points. Classical inverse Radon transform can be used in this scenario to obtain a 3D image, but using the Maximum Likelihood absorption tomography algorithm [54] further enhances the quality of the tomography reconstruction and performs well even for a small range of projections angels. In fact, advanced tomographic reconstruction algorithms based on the Maximum Likelihood principle are more resistant to noise and require fewer acquisitions, for a given precision, in comparison to standard tomographic protocols. We are investigating special tools to deal with informationally incomplete detection schemes for very high resolution and optimal methods for data analysis based on convex programming tools.

4.3. Quantum Tomography and Quantum Fisher Information

A further quantum approach to image analysis and detection schemes will be employed to achieve super-resolution (or, eventually, for maintaining the desired resolution and speeding up acquisition and elaboration of the quantum plenoptic images) and to compare correlation plenoptic detection scheme to the ultimate quantum limits. The basic concept underpinning the Fisher Information super-resolution imaging is the formal mathematical analogy between the classical wave optics and quantum theory, which makes it possible to apply the advanced tools of quantum detection and estimation theory to classical imaging and metrology [29,55]. The advantage of this approach lies in the ability to quantify the performance of the imaging setup based on rigorous statistical quantities. Inspired by the quantum theory of detection and estimation, quantum Fisher information, a quantity connected with the ultimate limits allowed by nature, is computed for simple 3D imaging scenarios like localization and resolution of two points in the object 3D space [30,56–58]. For example, one might be interested in measuring the separation of two point-like sources and seeking the optimal detection scheme, extracting the maximum amount of information about this parameter. We shall thus employ quantum Fisher information to design optimal measurement protocols within the quantum plenoptic devices, able to extract specific relevant information, for enhanced resolution, with a minimal number of acquisitions. The question of the minimal number of detections over the correlation planes which achieves acceptable reconstruction quality is of relevance because of the direct connection to the amount of processed data and can lead to reduced time for data processing.

5. Perspectives

In the context of the Qu3D project, all the developments and technologies presented in the previous sections will be integrated into the implementation of two quantum plenoptic imaging (QPI) devices, namely,

- a compact single-lens plenoptic camera for 3D imaging, based on the photon number correlations of a dim chaotic light source;
- an ultra-low noise plenoptic device, based on the correlation properties of entangled photon pairs emitted by spontaneous parametric down-conversion (SPDC), enabling 3D imaging of low-absorbing samples, at the shot-noise limit or below.

Our objective is to achieve, in both devices, high resolution, whether diffraction-limited or sub-Rayleigh, combined with a DOF larger by even one order of magnitude compared to standard imaging. The science and technology developed in the project will contribute to establishing a solid baseline of knowledge and skills for the development of a new generation of imaging devices, from quantum digital cameras enhanced by refocusing capability to quantum 3D microscopes [37] and space imaging devices.

6. Conclusions

We have presented the challenging research directions we are following to achieve practical quantum 3D imaging: minimizing the acquisition speed without renouncing to high SNR, high resolution, and large DOF [1–5]. Our work represents a significant advance with respect to the state-of-the-art of both classical and quantum imaging, as it enhances the performances of plenoptic imaging and dramatically speeds up quantum imaging, thus facilitating the real-world deployment of quantum plenoptic cameras.

This ambitious goal will be facilitated by working toward the extension of the reach of quantum imaging to other fields of science, and opening the way to new opportunities and applications, including the prospects of offering new medical diagnostic tools, such as 3D microscopes for biomedical imaging, as well as novel devices (quantum digital cameras enhanced by 3D imaging, refocusing, and distance detection capabilities) for security, space imaging, and industrial inspection. The collaboration crosses the traditional boundaries between the involved disciplines: quantum imaging, ultra-fast cameras, low-level programming of GPU, compressive sensing, quantum information theory, and signal processing.

Author Contributions: Conceptualization, M.D., Z.H., J.Ř., S.B., P.M., E.C. and C.B.; methodology, F.V.P., G.M., M.P. (Martin Paur), B.S., S.B., P.M., E.C. and C.B.; software, F.S. (Francesco Scattarella), G.M., L.M., M.P. (Michal Peterek), M.I. (Michele Iacobellis), I.P., F.S. (Francesca Santoro), P.M., A.U. and M.W.; firmware, P.M., A.U. and M.W.; validation, F.D.L., G.M., M.I. (Michele Iacobellis), P.M., A.U. and M.W.; formal analysis, F.D.L., M.I. (Michele Iacobellis), I.P. and F.S. (Francesca Santoro); setup implementation, P.M., A.U. and M.W.; setup design, F.D.L., D.G. and S.V.; writing—original draft preparation, F.V.P., I.P., F.S. (Francesca Santoro), S.B. and C.B.; writing—review and editing, F.D.L., F.V.P., I.P., F.S. (Francesca Santoro) and C.B.; supervision, M.D., A.G., F.V.P., C.A., L.A., E.C. and C.B.; project administration, M.D., C.A. and L.A.; funding acquisition, M.D., C.A., M.I. (Maria Ieronymaki) and C.B. All authors have read and agreed to the published version of the manuscript.

Funding: Project Qu3D is supported by the Italian Istituto Nazionale di Fisica Nucleare, the Swiss National Science Foundation (grant 20QT21_187716 "Quantum 3D Imaging at high speed and high resolution"), the Greek General Secretariat for Research and Technology, the Czech Ministry of Education, Youth and Sports, under the QuantERA program, which has received funding from the European Union's Horizon 2020 research and innovation program.

Informed Consent Statement: Not applicable.

Data Availability Statement: Relevant data are available from the authors upon reasonable request.

Conflicts of Interest: The authors declare no conflict of interest.

Abbreviations

The following abbreviations are used in this manuscript:

SPAD	Single-Photon Avalanche Diode
QPI	Quantum Plenoptic Imaging
fps	frames per second
GPU	Graphics Processing Unit
CPI	Correlation Plenoptic Imaging
SNR	Signal-to-Noise Ratio

Appl. Sci. **2021**, *11*, 6414

DOF	Depth of Field
EPR	Einstein–Podolski–Rosen
FPGA	Field-Programmable Gate Array
CPU	Central Processing Unit
GI	Ghost Imaging
CS	Compressive Sensing
DCT	Discrete Cosine Transform

References

1. Sansoni, G.; Trebeschi, M.; Docchio, F. State-of-The-Art and Applications of 3D Imaging Sensors in Industry, Cultural Heritage, Medicine, and Criminal Investigation. *Sensors* **2009**, *9*, 568. [CrossRef]
2. Geng, J. Structured-light 3D surface imaging: A tutorial. *Adv. Opt. Photonics* **2011**, *3*, 128. [CrossRef]
3. Hansard, M.; Lee, S.; Choi, O.; Horaud, R. *Time of Flight Cameras: Principles, Methods, and Applications*, 2013th ed.; Springer: Berlin, Germany, 2013.
4. Mertz, J. *Introduction to Optical Microscopy*; Cambridge University Press: Cambridge, UK, 2019.
5. Prevedel, R.; Yoon, Y.-G.; Hoffmann, M.; Pak, N.; Wetzstein, G.; Kato, S.; Schrödel, T.; Raskar, R.; Zimmer, M.; Boyden, E.S.; et al. Simultaneous whole-animal 3D imaging of neuronal activity using light-field microscopy. *Nat. Methods* **2014**, *11*, 727. [CrossRef]
6. Hall, E.M.; Thurow, B.S.; Guildenbecher, D.R. Comparison of three-dimensional particle tracking and sizing using plenoptic imaging and digital in-line holography. *Appl. Opt.* **2016**, *55*, 6410–6420. [CrossRef]
7. Kim, M.K. Principles and techniques of digital holographic microscopy. *SPIE Rev.* **2010**, *1*, 018005. [CrossRef]
8. Zheng, G.; Horstmeyer, R.; Yang, C. Wide-field, high-resolution Fourier ptychographic microscopy. *Nat. Photonics* **2013**, *7*, 739. [CrossRef] [PubMed]
9. Albota, M.A.; Aull, B.F.; Fouche, D.G.; Heinrichs, R.M.; Kocher, D.G.; Marino, R.M.; Mooney, J.G.; Newbury, N.R.; O'Brien, M.E.; Player, B.E.; et al. Three-dimensional imaging laser radars with Geiger-mode avalanche photodiode arrays. *Lincoln Lab. J.* **2002**, *13*, 351–370.
10. Marino, R.M.; Davis, W.R. Jigsaw: A foliage-penetrating 3D imaging laser radar system. *Lincoln Lab. J.* **2005**, *15*, 23–36.
11. Hansard, M.; Lee, S.; Choi, O.; Horaud, R.P. *Time-of-Flight Cameras: Principles, Methods and Applications*; Springer Science & Business Media: Berlin, Germany, 2012.
12. McCarthy, A.; Krichel, N.J.; Gemmell, N.R.; Ren, X.; Tanner, M.G.; Dorenbos, S.N.; Zwiller, V.; Hadfield, R.H.; Buller, G.S. Kilometer-range, high resolution depth imaging via 1560 nm wavelength single-photon detection. *Opt. Express* **2013**, *21*, 8904–8915. [CrossRef] [PubMed]
13. McCarthy, A.; Ren, X.; Della Frera, A.; Gemmell, N.R.; Krichel, N.J.; Scarcella, C.; Ruggeri, A.; Tosi, A.; Buller, G.S. Kilometer-range depth imaging at 1550 nm wavelength using an InGaAs/InP single-photon avalanche diode detector. *Opt. Express* **2013**, *21*, 22098–22113. [CrossRef]
14. Altmann, Y.; McLaughlin, S.; Padgett, M.J.; Goyal, V.K.; Hero, A.O.; Faccio, D. Quantum-inspired computational imaging. *Science* **2018**, *361*, eaat2298. [CrossRef] [PubMed]
15. Mertz, J. *Introduction to Optical Microscopy*; Roberts and Company Publishers: Englewood, CO, USA, 2010; Volume 138.
16. Pepe, F.V.; Di Lena, F.; Mazzilli, A.; Garuccio, A.; Scarcelli, G.; D'Angelo, M. Diffraction-limited plenoptic imaging with correlated light. *Phys. Rev. Lett.* **2017**, *119*, 243602. [CrossRef] [PubMed]
17. Zappa, F.; Tisa, S.; Tosi, A.; Cova, S. Principles and features of single-photon avalanche diode arrays. *Sens. Actuators A* **2007**, *140*, 103–112. [CrossRef]
18. Charbon, E. Single-photon imaging in complementary metal oxide semiconductor processes. *Philos. Trans. R. Soc. A* **2014**, *372*, 20130100. [CrossRef] [PubMed]
19. Antolovic, I.M.; Bruschini, C.; Charbon, E. Dynamic range extension for photon counting arrays. *Opt. Express* **2018**, *26*, 22234. [CrossRef] [PubMed]
20. Veerappan, C.; Charbon, E. A low dark count p-i-n diode based SPAD in CMOS technology. *IEEE Trans. Electron Devices* **2016**, *63*, 65. [CrossRef]
21. Antolovic, I.M.; Ulku, A.C.; Kizilkan, E.; Lindner, S.; Zanella, F.; Ferrini, R.; Schnieper, M.; Charbon, E.; Bruschini, C. Optical-stack optimization for improved SPAD photon detection efficiency. *Proc. SPIE* **2019**, *10926*, 359–365.
22. Ulku, A.C.; Bruschini, C.; Antolovic, I.M.; Charbon, E.; Kuo, Y.; Ankri, R.; Weiss, S.; Michalet, X. A 512 × 512 SPAD image sensor with integrated gating for widefield FLIM. *IEEE J. Sel. Top. Quantum Electron.* **2019** *25*, 6801212. [CrossRef]
23. Nguyen, A.H.; Pickering, M.; Lambert, A. The FPGA implementation of an image registration algorithm using binary images. *J. Real Time Image Pr.* **2016**, *11*, 799. [CrossRef]
24. Holloway, J.; Kannan, V.; Zhang, Y.; Chandler, D.M.; Sohoni, S. GPU Acceleration of the Most Apparent Distortion Image Quality Assessment Algorithm. *J. Imaging* **2018**, *4*, 111. [CrossRef]
25. Dadkhah, M.; Deen, M.J.; Shirani, S. Compressive Sensing Image Sensors-Hardware Implementation. *Sensors* **2013**, *13*, 4961. [CrossRef]
26. Chan, S.H.; Elgendy, O.A.; Wang, X. Images from Bits: Non-Iterative Image Reconstruction for Quanta Image Sensors. *Sensors* **2016**, *16*, 1961. [CrossRef]

27. Rontani, D.; Choi, D.; Chang, C.-Y.; Locquet, A.; Citrin, D.S. Compressive Sensing with Optical Chaos. *Sci. Rep.* **2016**, *6*, 35206. [CrossRef]
28. Gul, M.S.K.; Gunturk, B.K. Spatial and Angular Resolution Enhancement of Light Fields Using Convolutional Neural Networks. *IEEE Trans. Image Process.* **2018**, *27*, 2146. [CrossRef]
29. Motka, L.; Stoklasa, B.; D'Angelo, M.; Facchi, P.; Garuccio, A.; Hradil, Z.; Pascazio, S.; Pepe, F.V.; Teo, Y.S.; Rehacek, J.; et al. Optical resolution from Fisher information. *EPJ Plus* **2016**, *131*, 130. [CrossRef]
30. Řeháček, J.; Paúr, M.; Stoklasa, B.; Koutný, D.; Hradil, Z.; Sánchez-Soto, L.L. Intensity-based axial localization at the quantum limit. *Phys. Rev. Lett.* **2019**, *123*, 193601. [CrossRef] [PubMed]
31. QuantERA Call 2019. Available online: https://www.quantera.eu/calls-for-proposals/call-2019 (accessed on 22 March 2021).
32. Raytrix. 3D Light Field Camera Technology. Available online: https://raytrix.de (accessed on 22 March 2021).
33. D'Angelo, M.; Pepe, F.V.; Garuccio, A.; Scarcelli, G. Correlation Plenoptic Imaging. *Phys. Rev. Lett.* **2016**, *116*, 223602. [CrossRef] [PubMed]
34. Pepe, F.V.; Di Lena, F.; Garuccio, A.; Scarcelli, G.; D'Angelo, M. Correlation Plenoptic Imaging with Entangled Photons. *Technologies* **2016**, *4*, 17. [CrossRef]
35. Pepe, F.V.; Vaccarelli, O.; Garuccio, A.; Scarcelli, G.; D'Angelo, M. Exploring plenoptic properties of correlation imaging with chaotic light. *J. Opt.* **2017**, *19*, 114001. [CrossRef]
36. Di Lena, F.; Massaro, G.; Lupo, A.; Garuccio, A.; Pepe, F.V.; D'Angelo, M. Correlation plenoptic imaging between arbitrary planes. *Opt. Express* **2020**, *28*, 35857. [CrossRef] [PubMed]
37. Scagliola, A.; Di Lena, F.; Garuccio, A.; D'Angelo, M.; Pepe, F.V. Correlation plenoptic imaging for microscopy applications. *Phys. Lett. A* **2020**, *384*, 126472. [CrossRef]
38. D'Angelo, M.; Shih, Y.H. Quantum Imaging. *Laser Phys. Lett.* **2005**, *2*, 567—596. [CrossRef]
39. Gatti, A.; Brambilla, E.; Bache, M.; Lugiato, L.A. Ghost Imaging with Thermal Light: Comparing Entanglement and Classical Correlation. *Phys. Rev. Lett.* **2004**, *93*, 093602. [CrossRef] [PubMed]
40. Scala, G.; D'Angelo, M.; Garuccio, A.; Pascazio, S.; Pepe, F.V. Signal-to-noise properties of correlation plenoptic imaging with chaotic light. *Phys. Rev. A* **2019**, *99*, 053808. [CrossRef]
41. Brida, G.; Genovese, M.; Ruo Berchera, I. Experimental realization of sub-shot-noise quantum imaging. *Nat. Photonics* **2010**, *4*, 227. [CrossRef]
42. Ferri, F.; Magatti, D.; Lugiato, L.A.; Gatti, A. Differential Ghost Imaging. *Phys. Rev. Lett* **2010**, *104*, 253603. [CrossRef] [PubMed]
43. Bruschini, C.; Homulle, H.; Antolovic, I.M.; Burri, S.; Charbon, E. Single-photon avalanche diode imagers in biophotonics: Review and outlook. *Light Sci. Appl.* **2019**, *8*, 87. [CrossRef]
44. Caccia, M.; Nardo, L.; Santoro, R.; Schaffhauser, D. Silicon photomultipliers and SPAD imagers in biophotonics: Advances and perspectives. *Nucl. Instrum. Methods Phys. Res. A* **2019**, *926*, 101–117. [CrossRef]
45. Ulku, A.; Ardelean, A.; Antolovic, M.; Weiss, S.; Charbon, E.; Bruschini, C.; Michalet, X. Wide-field time-gated SPAD imager for phasor-based FLIM applications. *Methods Appl. Fluoresc.* **2020**, *8*, 024002. [CrossRef]
46. Zanddizari, H.; Rajan, S.; Zarrabi, H. Increasing the quality of reconstructed signal in compressive sensing utilizing Kronecker technique. *Biomed. Eng. Lett.* **2018**, *8*, 239. [CrossRef]
47. Mertens, L.; Sonnleitner, M.; Leach, J.; Agnew, M.; Padgett, M.J. Image reconstruction from photon sparse data. *Sci. Rep.* **2017** *7*, 42164. [CrossRef]
48. Katz, O.; Bromberg, Y.; Silberberg, Y. Compressive ghost imaging. *Appl. Phys. Lett.* **2009**, *95*, 131110. [CrossRef]
49. Jiying, L.; Jubo, Z.; Chuan, L.; Shisheng, H. High-quality quantum-imaging algorithm and experiment based on compressive sensing. *Opt. Lett.* **2010**, *35*, 1206–1208. [CrossRef] [PubMed]
50. Aßmann, M.; Bayer, M. Compressive adaptive computational ghost imaging. *Sci. Rep.* **2013**, *3*, 1545.
51. Chen, Y.; Cheng, Z.; Fan, X.; Cheng, Y.; Liang, Z. Compressive sensing ghost imaging based on image gradient. *Optik* **2019**, *182*, 1021–1029.
52. Liu, H.-C. Imaging reconstruction comparison of different ghost imaging algorithms. *Sci. Rep.* **2020**, *10*, 14626. [CrossRef] [PubMed]
53. Tibshirani, R. Regression shrinkage and selection via the lasso: A retrospective. *J. R. Statist. Soc. B* **2011**, *73*, 273–282. [CrossRef]
54. Řeháček, J.; Hradil, Z.; Zawisky, M.; Treimer, W.; Strobl, M. Maximum-likelihood absorption tomography. *EPL* **2002**, *59*, 694. [CrossRef]
55. Hradil, Z.; Rehacek, J.; Fiurasek, J.; Jezek, M. Maximum Likelihood Methods in Quantum Mechanics, in Quantum State Estimation. In *Lecture Notes in Physics*; Paris, M.G.A., Rehacek, J., Eds.; Springer: Berlin, Germany, 2004; pp. 59–112.
56. Rehacek, J.; Hradil, Z.; Stoklasa, B.; Paur, M.; Grover, J.; Krzic, A.; Sanchez-Soto, L.L. Multiparameter quantum metrology of incoherent point sources: Towards realistic superresolution. *Phys. Rev. A* **2017**, *96*, 062107. [CrossRef]
57. Paur, M.; Stoklasa, B.; Hradil, Z.; Sanchez-Soto, L.L.; Rehacek, J. Achieving the ultimate optical resolution. *Optica* **2016**, *3*, 1144. [CrossRef]
58. Paur, M.; Stoklasa, B.; Grover, J.; Krzic, A.; Sanchez-Soto, L.L.; Hradil, Z.; Rehacek, J. Tempering Rayleigh's curse with PSF shaping. *Optica* **2018**, *5*, 1177. [CrossRef]

MDPI

St. Alban-Anlage 66

4052 Basel

Switzerland

Tel. +41 61 683 77 34

Fax +41 61 302 89 18

www.mdpi.com

Applied Sciences Editorial Office

E-mail: applsci@mdpi.com

www.mdpi.com/journal/applsci